THERMAL ENERGY STORAGE

From Fundamentals
to Applications

THERMAL ENERGY STORAGE

From Fundamentals to Applications

Alexandra Soh
Vivekh Prabakaran
Ernest Kian Jon Chua
National University of Singapore, Singapore

World Scientific

NEW JERSEY · LONDON · SINGAPORE · BEIJING · SHANGHAI · HONG KONG · TAIPEI · CHENNAI · TOKYO

Published by

World Scientific Publishing Co. Pte. Ltd.

5 Toh Tuck Link, Singapore 596224

USA office: 27 Warren Street, Suite 401-402, Hackensack, NJ 07601

UK office: 57 Shelton Street, Covent Garden, London WC2H 9HE

Library of Congress Cataloging-in-Publication Data
Names: Soh, Alexandra, author. | Prabakaran, Vivekh, author. | Chua, Ernest Kian Jon, author.
Title: Thermal energy storage : from fundamentals to applications / Alexandra Soh,
    Vivekh Prabakaran, Ernest Kian Jon Chua, National University of Singapore, Singapore.
Description: USA : World Scientific, [2023] | Includes bibliographical references and index.
Identifiers: LCCN 2022058131 | ISBN 9789811271175 (hardcover) |
    ISBN 9789811271182 (ebook for institutions) | ISBN 9789811271199 (ebook for individuals)
Subjects: LCSH: Heat storage.
Classification: LCC TJ260 .S64 2023 | DDC 621.47/12--dc23/eng/20230111
LC record available at https://lccn.loc.gov/2022058131

British Library Cataloguing-in-Publication Data
A catalogue record for this book is available from the British Library.

For any available supplementary material, please visit
https://www.worldscientific.com/worldscibooks/10.1142/13276#t=suppl

Desk Editors: Nimal Koliyat/Steven Patt

Typeset by Stallion Press
Email: enquiries@stallionpress.com

Preface

Thermal Energy Storage (TES) systems play an essential role in the energy distribution landscape of today's world. As industry and academia move towards more efficient methods in balancing supply and demand in thermal applications, advancements in TES will continue to feature more prominently in the conception of novel energy distribution systems. TES is a technology that stocks thermal energy either by heating or cooling a storage medium so that the stored energy can be released and used later for heating and cooling applications or even for power generation. Often, TES systems operate in tandem with renewable thermal technologies, which harness heat from a wide range of local renewable sources, including solar, geothermal, and industrial waste heat. TES systems are used particularly in buildings and in industrial processes.

This book presents comprehensive content on the key concepts underpinning the field of TES systems and approaches in terms of both their characterization and optimization. Analysis and experimental techniques are presented to facilitate a better understanding of the design, handling, and operation of such systems. Recent advancements in academia and industrial applications, including economic analyses, are further described to highlight the current direction of innovation and trends in the field.

The contents of this monograph are specifically arranged into six essential chapters. The first chapter provides a comprehensive introduction to TES systems. The benefits and drawbacks of TES systems are discussed. Several applications of TES in the industrial landscape are explored, particularly in urban cooling/heating. It also covers key heat transfer and thermodynamic concepts on TES and briefly introduces the different types of TES.

In the second chapter, the primary concepts of sensible thermal energy storage systems are presented with a strong emphasis on the most common types of sensible thermal energy storage systems in both urban and commercial contexts. It further provides key thermodynamic principles and characterization techniques and describes the many different forms of sensible thermal energy storage.

The third chapter focuses on latent thermal energy storage systems and the wide range of characterization and analysis approaches. The concept of latent heat and phase-change materials is first introduced, along with the key thermodynamic and heat transfer concepts. In addition, it includes the experimental methods for the derivation and analyses of PCM properties, highlighting their respective benefits and drawbacks. Also, the application of PCMs in thermal energy storage systems is discussed, and the different ways they are applied (encapsulated, unencapsulated, building façade, etc.) are reviewed. Finally, the end of the chapter leverages case studies to illustrate the wide range of applications of PCM latent thermal storage systems.

The fourth chapter documents the details of thermochemical energy storage. It provides the key concepts behind thermochemical energy storage and some modelling techniques to simulate and predict its performance.

The fifth chapter entails the method to design a TES system for a specific application. This chapter describes the basic principles of designing general TES and the variety of ways in which thermal energy storage can be retrofitted into an existing system. Essential operating and maintenance considerations are also elaborated.

Chapter six, the last chapter, focuses on recent key developments in TES designs tailored for specific applications in different industrial sectors, including industrial buildings, liquefied natural gas (LNG) terminals, and district cooling plants. Several case studies are provided to demonstrate the applications of both sensible TES and latent TES for large-scale systems incorporating some levels of life-cycle economic analyses.

The chapters are judiciously selected to balance fundamental concepts, industrial applications, and recent research developments. It is worthy to note that the content includes recent fundamental advances in sensible TES and latent TES and demonstrates the employment of TES in several niche industrial applications that are currently generating much interest due to global interest in energy sustainability. In sum, this book serves as a one-stop archive that connects recent fundamentals and developments in TES to applications. Pertinent materials have been carefully selected from the literature and our published works. Credits should belong to the original sources.

To express gratitude and appreciation, the authors would like to extend their heartfelt thanks to team members who have assisted and contributed to the documentation of the technical content presented in the various chapters. Some of these include research staff, ex-PhD students, and other graduate students who have worked and contributed during their residence in our laboratories at different points in time.

About the Authors

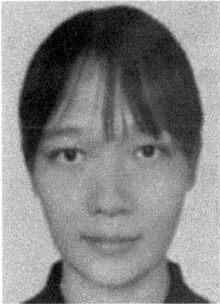

Alexandra Soh is currently a Research Engineer with the Department of Mechanical Engineering, National University of Singapore (NUS), and is completing her PhD on low-temperature thermal energy storage for cold recovery from LNG regasification. Having obtained her bachelor's degree in mechanical engineering from the National University of Singapore in 2015, she joined the Energy Research Institute @ NTU (ERI@N) for 3 years as a research associate, overseeing a wide range of industrial research projects primarily focused on novel sustainable building technologies. After completing her master's degree with the Nanyang Technological University in 2018, Alexandra returned to NUS to pursue her PhD. Her key research foci include latent thermal energy storage systems, novel cooling technologies, thermal systems design, and multi-objective optimization. Expertise: Thermal energy storage, air-conditioning and dehumidification, district cooling systems, thermal systems.

Vivekh Prabakaran currently works as a Postdoctoral Research Fellow with the Department of Mechanical Engineering, National University of Singapore (NUS). He completed his undergraduate coursework from BITS Pilani, India (2015), and worked as a research assistant at the Heat Transfer Laboratory of the Indian Institute of Science. Being the recipient of the prestigious President's Graduate Fellowship, he received his PhD in Mechanical Engineering from NUS in 2020 under Professor Kian Jon Chua's supervision. During his doctoral training, he synthesized next-generation advanced desiccant materials and conducted key fundamental experimental and theoretical investigations on heat and mass transport. His PhD findings have been published in over 7 peer-reviewed international journals. Vivekh's specific research focus areas include dehumidification, cooling, heat pumping, desalination, solar energy, multi-criteria decision analysis, and advanced materials for energy research. Research interests: Dehumidification, new materials for energy research, air-conditioning, optimization, desalination, solar energy.

Ernest Kian Jon Chua is currently an Associate Professor with the Department of Mechanical Engineering, National University of Singapore (NUS). He has been conducting research on cooling, dehumidification, and thermal energy processes since 2001. He has conducted both modelling and experimental works for specific thermal energy systems. He is highly skilled in designing, fabricating, commissioning, and testing many sustainable energy systems to provide heating, cooling, and humidity control for both small- and large-scale applications. He has been elected to several fellowships including Fellow of Royal Society, Fellow of Energy Institute, and Fellow of IMechE. He has more than 200 international peer-reviewed journal publications, 6 book chapters, and 2 recent monographs on advances in air-conditioning (https://www.springer.com/gp/book/9789811584763 and https://www.springer.com/gp/book/9783030808426). He is ranked first out of 500 worldwide scholars on air-conditioning and evaporative cooling research (2012 to 2021) based on Scopus SciVal topic cluster. He was highlighted among the top 1% of scientists in the world by the Universal Scientific Education and Research Network and top 0.5% in the Stanford list of energy researchers. Based on Google Scholar, his works have garnered more than 12,300 citations with a current H-index of 58. Further, he owns more than 10 patents related to several innovative cooling and dehumidification systems. He has been invited to deliver more than 30+ plenary/keynote/invited talks over the last 3–4 years at international (Europe, UK, and US) conferences on innovative technology, energy, and sustainability. He is the Principal Investigator of several multi-million competitive research grants. Additionally, he has been awarded multiple local, regional, and international awards for his breakthrough research endeavours. His research has appeared in many scientific forums and (inter)national media, including Bloomberg, ASME, Channel News Asia, etc. Expertise: Air-conditioning, dehumidification, refrigeration, district cooling, cogeneration/tri-generation/quad-generation, thermal energy processes and systems.

Contents

Chapter 1

Introduction to Thermal Energy Storage Systems

Abstract

The practice of storing thermal energy dates back to ancient civilizations from forms such as storage of ice blocks buried in sawdust and straw, to the use of heated rocks for cooking and warmth in colder climates. Modern-day thermal energy storage, in contrast, has taken on much more sophisticated forms in both domestic and commercial settings, which have become the mainstays in thermal system designs around the world. While simplistic in concept, the nuances in application and design have spawned a wide range of in-depth research, which has led to a rich and diverse field of study in society today. This opening chapter intends on providing a brief overview of the role of thermal energy storage in today's world. Additionally, the classifications of thermal energy storage will be covered without going into excessive detail. Lastly and most importantly, the various thrusts of research and development today will be discussed.

Keywords: Thermal energy storage systems; process integration.

1.1. Introduction

An energy-efficient thermal energy storage (TES) system constitutes a crucial instrument in thermal distribution networks today. As the thermal management of urban and industrial landscapes becomes increasingly important, so too does the impact of carefully planned and designed TES systems. TES systems, for the most part, do not specifically generate nor consume thermal energy — their role in the thermal distribution landscape is essentially analogous to that of a battery or capacitor in a power distribution network. They allow supply to meet demand by bridging the gaps between the two. In most manifestations of TES systems, this gap is often a time gap, where the surplus thermal energy generated (or removed, for cooling applications) is preserved in the TES for a future period when the demand for it is greater or more appropriate, such as with hot water storage for night-time heating, or even across longer periods of time like seasonal TES systems which store heat during summer and discharge during winter. The "storage" phase of the TES is often coined as the "charging" phase in the literature, while the utility phase of the TES process is considered as the "discharging phase". Hence, TES systems play important roles in creating efficient and resilient thermal distribution networks by allowing *load matching* to take place as needed. Figure 1.1 illustrates how TES is applied to bridge the gap between demand and supply needs, while Figure 1.2 provides some examples of existing well-known TES systems.

Applications of TES can be broadly classified based on their respective operating temperature ranges. While this method of classification can be considered largely arbitrary, the requirements of the TES system and the application that it serves are generally well

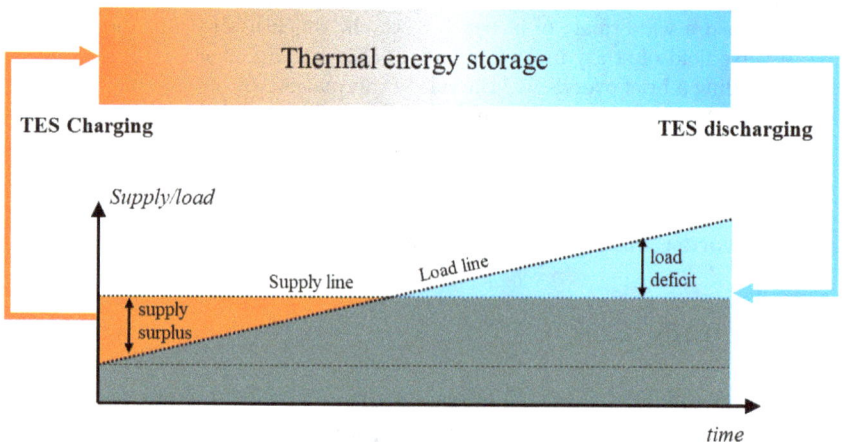

Figure 1.1.   TES allows for surplus generation to meet load deficits by bridging the time gap between supply-side and demand-side utilities.

(a)

(b)

(c)

Figure 1.2. (a) Molten salt thermal energy storage tank in a concentrating solar plant [1], (b) phase-change material (PCM) encapsulated in high-density polyethylene sphere [2], and (c) ice-on-coil thermal storage systems for cooling applications [3].

understood based solely on its operating temperature conditions. For simplicity, TES systems are often classified into *high* temperature and *low* temperature applications as depicted in Figure 1.3.

Guelpa and Verda [4] described the key benefits and drawbacks of TES systems in thermal distribution networks as such:

- Advantages of TES systems
    i. Increased flexibility in operation;
    ii. Allows for downsizing of chillers/boilers, leading to a reduction in capital investment costs and carbon footprint;

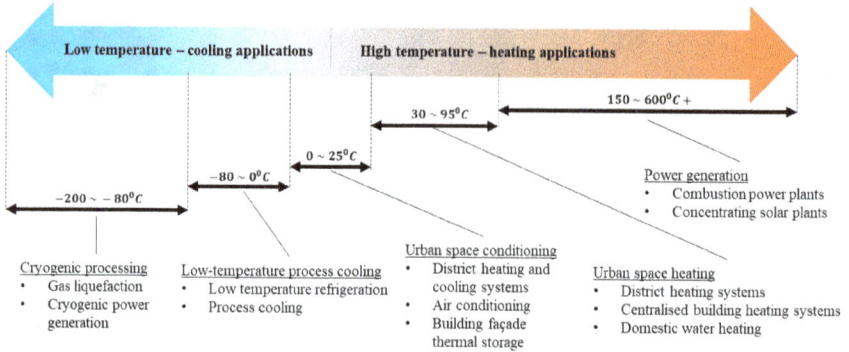

Figure 1.3.   Common TES applications over the various operating temperature ranges.

    iii. Improves overall energy and cost efficiency of the system, allowing for economic cost and carbon emission savings; and

    iv. Smoothens out supply intermittencies in renewable energy sources.

- Drawbacks of TES systems

    i. Can be costly depending on the type of TES chosen;

    ii. Space investment is high for low-energy density TES systems;

    iii. Thermal losses in storage and additional piping can be significant and negatively impact the cost-efficiency of the installation; and

    iv. Some aspects of TES design and operation are still poorly understood and challenging to generalise.

In 2020, the International Renewable Energy Agency (IRENA) reported a total of 91 GWh of TES capacity for building heating and 105 GWh for district heating systems globally. For cooling applications, a total of 14 GWh TES capacity was estimated globally, and this figure is forecasted to increase to 26 GWh by 2030 [5]. As variable renewable energy is estimated to occupy more than 60% of power generation by 2050 (compared to 10% in 2018), energy storage flexibility is expected to become key for the future energy distribution landscape. The continued progress and integration of storage technologies, particularly TES technologies is, hence, expected to be essential and critical for future energy systems, as portrayed in Figure 1.4.

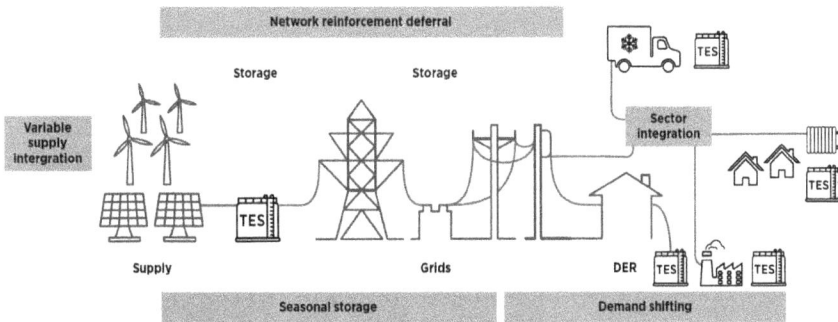

Figure 1.4. Integration of TES into the energy sector as envisioned by IRENA [5].

1.2. Classifications of Thermal Energy Storage Systems

TES systems are best classified based on the specific mechanics of their storage methods. Most known thermal energy storage systems such as domestic hot water storage systems and building chilled water storage systems are classified as sensible TES (STES) systems while other systems that capitalise on a change of phase of the storage media are considered *latent* TES (LTES). Among the least explored are the *thermochemical* TES (TTES) systems, which utilise reversible chemical reactions for the storage and delivery of heat. Comparatively, each thermal energy storage possesses its respective benefits and drawbacks, as well as their thermal energy densities relative to the others. Thermochemical TES, for example, have the highest thermal energy densities but is largely unexplored and wide-scale implementation is still considered to be in its infancy. In contrast, sensible and some latent TES systems are well applied commercially despite having lower energy densities, due to their ease of implementation. Figure 1.5 illustrates the classification of TES technologies based on their energy storage mechanics.

1.2.1. *Sensible thermal energy storage systems*

The key difference between STES systems and other TES systems is largely due to their thermal storage mechanics. STES systems are limited to the sensible heat gain arising from the specific heat capacitance of the storage media and its operating temperature range.

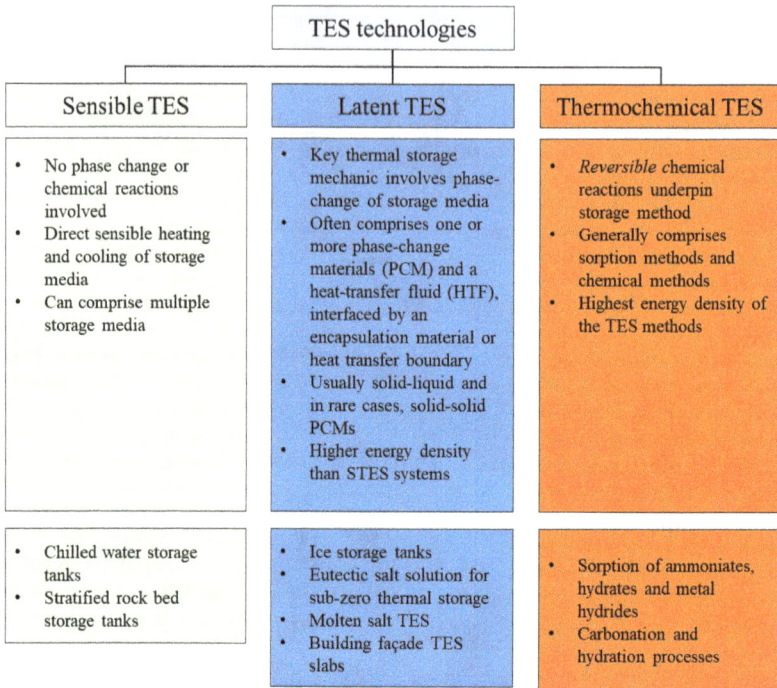

Figure 1.5.　Classification of TES methods incorporating examples.

Designated as being the most basic form of thermal energy storage, the energy density and performance of STES systems hinge strongly on the storage media involved, which can involve a single material or more. Commonly used single-media STES systems include hot and chilled water storage tanks for building space conditioning. In building-level and district-level cooling systems, excess chiller capacity particularly during off-peak hours is employed to charge chilled water storage tanks, which assist in partially or fully meeting the desired cooling demand during peak hours. This method of operation permits cost and energy savings. For instance, at lower off-peak electrical tariffs, the chiller operational cost is reduced for the same amount of cooling supplied. Further, operating the chiller under consistent full-load conditions means the potential of system inefficiencies arising from part-load conditions during off-peak hours is minimised; leading to energy savings. Additionally, cooling plants can downsize their chiller units; leading to further cost savings. Figure 1.6

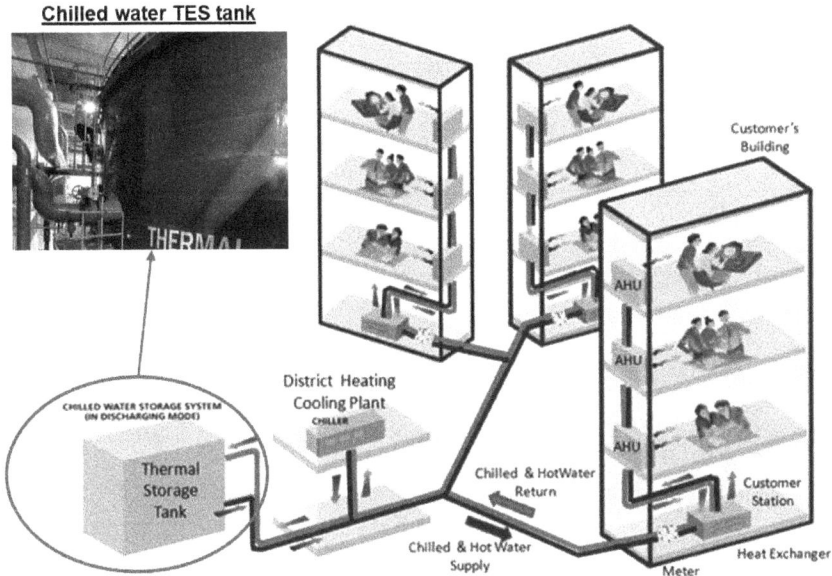

Figure 1.6.   Schematic of district heating and cooling system integrated with TES [6].

provides a schematic illustrating how TES can be incorporated into a district heating and cooling system.

Kocak *et al.* [7] provided an alternate perspective to classify STES systems as shown in Figure 1.6. Besides urban and domestic applications, STES has seen wide application in more advanced and larger scale contexts such as long-term seasonal TES systems, leading to TES technologies such as aquifer and borehole TES systems. Although some of the niche forms of TES depicted in Figure 1.7 will not be covered in this chapter, these examples serve as a reminder of the importance and prevalence of STES systems in today's energy landscape.

Despite having far lower thermal energy densities than LTES or TTES systems, STES systems excel in their simplicity of implementation and predictability of outcome, with less variability in thermal performance and control/monitoring methods. When there are no space constraints, STES systems are easily the most economical go-to option for TES systems. Figure 1.8 shows several examples of well-established long-term seasonal STES system types [6].

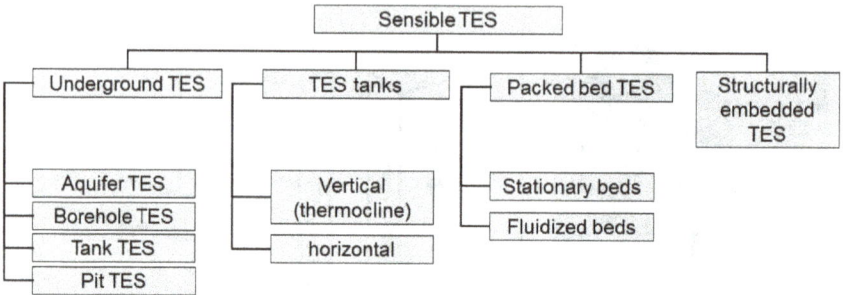

Figure 1.7.   Classification of STES systems [7].

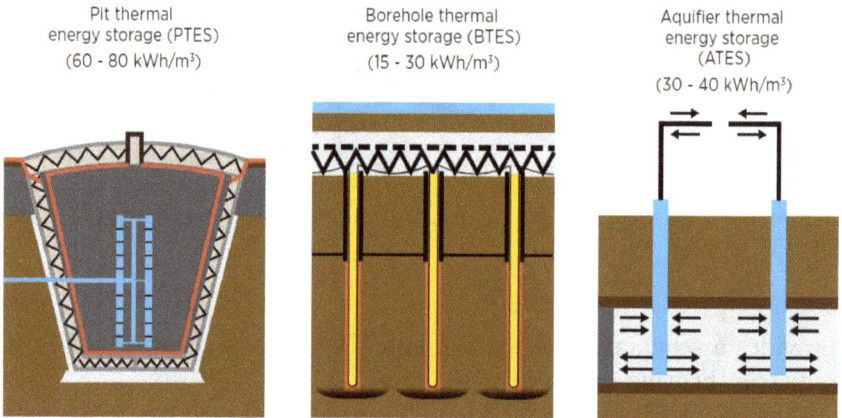

Figure 1.8.   Long-term seasonal STES systems [5].

1.2.2.   *Latent thermal energy storage systems*

Similar to STES systems, latent TES systems utilise the inherent thermal capacitance of the primary storage material used. However, unlike STES systems, the thermal storage capacity of an LTES system depends largely on the phase-change properties of its storage media which is the phase-change material (PCM). During the phase change process across the freezing or melting temperature range of a PCM, latent heat is either absorbed or released. As the latent heat of a material is many times greater than its specific heat, LTES systems allow for a significantly higher thermal storage capacity compared to STES for the same temperature range. This, however, does not

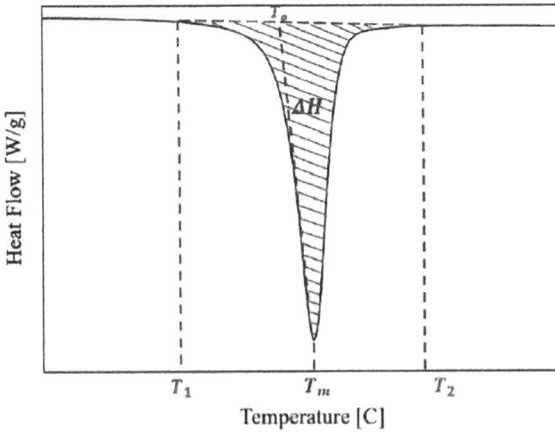

Figure 1.9. Latent heat during the phase-change process of a PCM as captured by the Differential Scanning Calorimetry method (DSC), represented by the area occupied under the curve. T_m represents the phase-change temperature of the PCM while T_1 and T_2 represent the onset temperatures of the melting and freezing temperatures respectively [8].

restrict LTES system operation to within the phase-change temperature range of the PCM. A significant portion of the thermal capacity of the system still relies on the sensible thermal capacitance of the PCM and other storage media contained in the LTES. Figure 1.9 illustrates the latent heat quantity during the phase-change process of a typical PCM.

Figure 1.10 demonstrates a simplified classification of the varying types of PCMs in LTES systems [8]. PCMs for LTES applications are also classified into several groups, the largest being the most commonly applied solid–liquid PCM, which relies on the solid–liquid phase change process to store thermal energy. Other less common LTES systems utilise PCMs such as solid–solid PCMs and liquid–gas PCMs, which are comparatively less established in industrial practice. For a typical solid–solid PCM, the material undergoes a reversible "phase" transition in the form of a solid structure, transforming between crystalline, semi-crystalline, or amorphous structures upon application or removal of thermal energy. Unlike solid–liquid PCMs, solid–solid PCMs tend to retain much of their bulk properties during the phase transition, making them exceedingly form-stable and durable for long-term application [9].

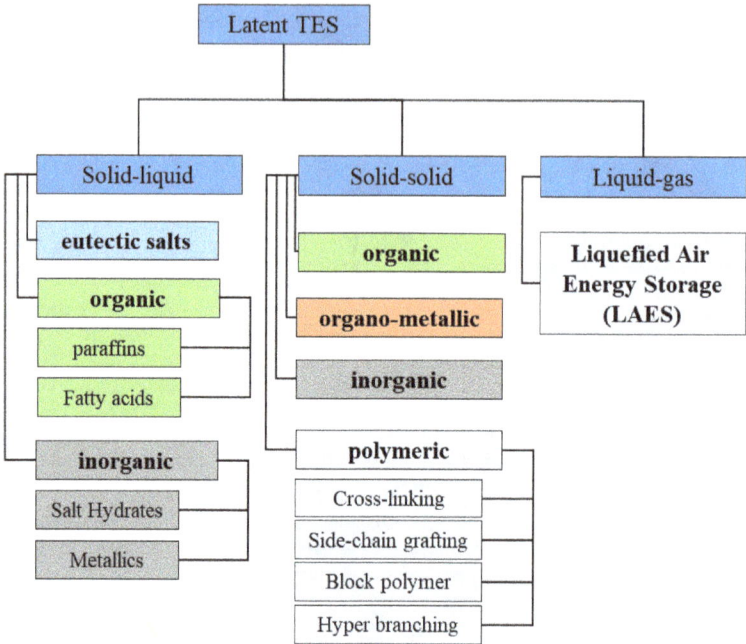

Figure 1.10.   Simplified classification of the varying types of PCMs in latent TES systems [9].

Figure 1.11 portrays the heat transfer process that occurs within an LTES, particularly between the HTF and the PCM across the interface. The PCM is unable to deliver thermal energy (hot or cold) during its solidified state, therefore, a separate heat-transfer fluid (HTF) within the system is required to perform the heat transfer between the LTES and its source or load. Two common approaches are available, namely, encapsulated PCM (EPCM) and non-encapsulated PCM LTES systems. Non-encapsulated systems constitute the more conventional approach in LTES systems where PCM are designed to occupy the majority of the LTES storage volume, with internal piping or heat exchange devices embedded in between the PCM space. This configuration maximises the LTES storage volume with as much PCM as could be reasonably contained within, at the cost of inconsistent and deteriorating heat transfer rates over greater levels of phase-change processes. Encapsulated systems, however, provide the PCM to be contained within receptacles of varying shapes and sizes, such as spheres, tubes, and slabs, with

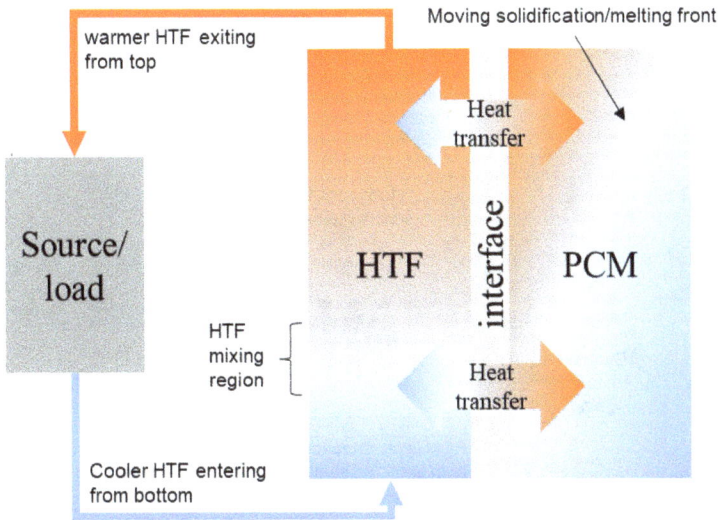

Figure 1.11. A generalised representation of an LTES system. Heat transfer within the LTES occurs primarily between the HTF and the PCM across the interface, which can take the form of internal tubes/heat exchanger surfaces in non-encapsulated systems or the receptacle shell in encapsulated systems. In stratified LTES systems, a mixing region between the hot and cold HTF will develop within the HTF, similar to that in stratified STES systems.

the HTF flowing on the exterior of the encapsulations. While this approach limits the amount of PCM that could be stored within the LTES itself, it provides key benefits from the perspective of better prediction of the heat transfer and stratification characteristics of the LTES.

In some encapsulated PCM LTES systems, however, an HTF is absent and the TES functions purely as a static thermal capacitor that operates between changes in ambient conditions. One widely piloted PCM-based LTES study involving this concept targets applications in building facades. Figure 1.12 illustrates the employment of the EPCM LTES specifically for building facades. Thermal management with building envelopes aims to moderate indoor thermal conditions within the temperature differences between night and day, especially in temperate climates where the conditions between night and day can vary greatly. Diurnal heat gain from the ambient is absorbed by the encapsulated melting PCM layer that is incorporated into the building façade during the day; moderating the

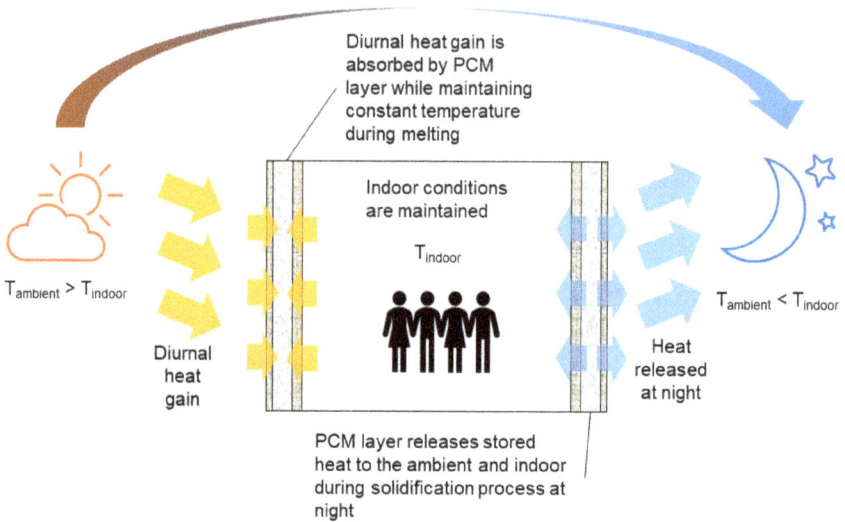

Figure 1.12.   EPCM LTES application for building facades. Diurnal heat gain in the day is absorbed by the PCM layer embedded in the building facade and released at night during the solidification process when ambient conditions cool below its phase-change temperature.

temperature of the indoor conditions. At night, the stored heat in the liquefied PCM is released to the ambient and indoor environment when the conditions fall below the PCM's phase-change temperature. Similarly, indoor conditions are moderated as heat is released during the solidification process to the indoors. While extensively studied in China and Europe, such technology has seen considerably less uptake by countries experiencing the tropical climate, where temperature ranges between day and night are considerably smaller [10].

While LTES systems possess highly favourable features in terms of improved overall energy density in contrast to STES systems, there are considerable constraints arising from the availability of commercial PCMs and their associated compatibilities with various applications (Figure 1.13). That is not every PCM yields the same degree of latent heat. Furthermore, operating conditions limit the types of PCMs that can be used, leading to further selection limitations during the design and sizing phases. Lastly, the added cost per unit mass of PCMs in comparison to widely available HTFs such as water or aqueous ethylene glycol solutions can lead to low-performing LTES

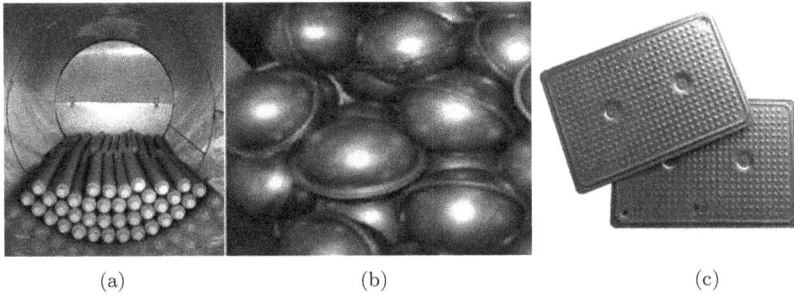

(a)   (b)   (c)

Figure 1.13.   (a) Commercial encapsulated PCM tubes; (b) spheres in TES tank; and (c) slab-type encapsulation [11,12].

designs, where the improvement over a STES system is insignificant for the additional cost.

The design and sizing of an LTES for its application and the selection of the required PCM is therefore a key area of many studies in literature, and recent advances in this aspect are presented in the ensuing sections.

1.2.3.   *Thermochemical thermal energy storage systems*

Thermochemical TES (TTES) systems, unlike STES and LTES systems, involve actual chemical reactions to store thermal energy. They are known to display high levels of thermal energy density, however, thus far, works have largely been confined to lab-scale. One key property of the TTES systems is that the thermal storage mostly occurs at ambient temperature [13], leading to its ability to store energy for longer periods of time with minimal losses. TTES systems are often loosely divided into two groups — sorption-based TTES or reaction-based TTES. Sorption-based TTES can be based on either absorption or adsorption. In the former, a solution at molecular level is formed between an absorbate (gas/liquid) and an absorbent (liquid/solid) material. Covalent bonding at the molecular scale results in a large amount of energy being stored or released. In the latter case, only surface interaction takes place between the adsorbent and substrate, where the substrate forms a cohesive boundary at the interaction surfaces with the adsorbent, while the internal composition of the adsorbent remains undisturbed. The adsorption process is also

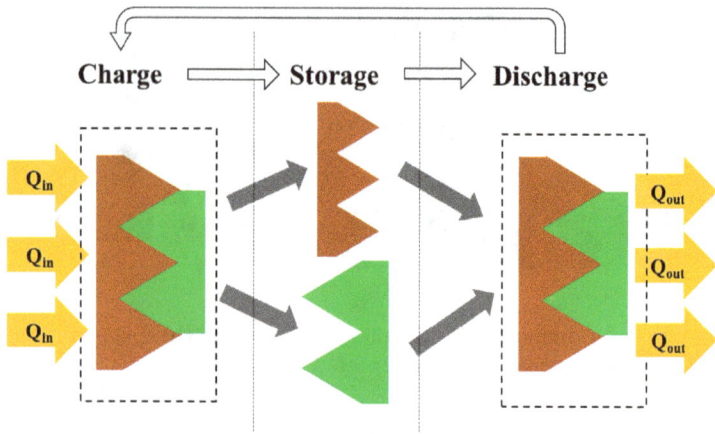

Figure 1.14.   The charge and discharge processes associated with sorption-type TTES systems. Q_{in} and Q_{out} refer to the charging heat input and discharging heat output, respectively. In air-conditioning applications, Q_{in} is more commonly known as the heat of regeneration while Q_{out} is known as the sorption heat.

distinguished between physisorption and chemisorption processes, wherein in the former, only Van der Waals interactions between the surface molecules of the adsorbent and the substrate take place, while actual chemical bonding occurs in the latter. While sorption processes are still in the preliminary stages of research within the field of TES, such sorption mechanics are already been well studied and deployed in the field of air-conditioning and dehumidification. Figure 1.14 depicts the charge and discharge processes associated with sorption-type TTES systems.

Figure 1.15 illustrates a few lab-scale sorption-based TTES for seasonal heat storage. Reaction-based TTES systems charge and discharge thermal energy by means of reversible endothermic and exothermic reactions. Thermal energy is absorbed during the endothermic process and discharged during the reverse exothermic process. Between the charge and discharge phases, the thermochemical materials (TCMs) are cooled to storage temperature and stored until discharging. As the energy released is dependent solely on the reaction kinetics and not the temperature state of the TCM constituents, the energy can be stored indefinitely with negligible losses. Key benefits of TTES systems include the wide availability of TCMs over large temperature ranges and minimal losses incurred over long

(a)   (b)

Figure 1.15. (a) Lab-scale sorption-based TTES for seasonal heat storage using a silica-gel pellet-based packed bed [14]; and (b) a finned-tube $SrBr_2/H_2O$ thermochemical energy storage heat exchanger system [15].

periods of storage time. A comparison between the three types of TES systems is detailed in Table 1.1.

1.3. Core Aspects of TES Study

The field of TES system study in academic literature can be broadly divided into three core foci: material, configuration, and system integration and control, as shown in Figure 1.16. Similar to other fields within the sphere of thermal engineering, the core thermal capacitance and reaction kinetics of the storage media significantly impact the storage capacity of the system. In addition, configuration and application of the material in the context of its use also constitutes a key area of understanding. This is particularly important for encapsulated PCM LTES systems, where the diversity of sizing and forms of encapsulation methods have far-reaching impacts on the heat transfer and stratification performance of the TES. Finally, systems integration, and the place of a TES within a thermal distribution network are other key factors. As the core function of a TES is fundamentally targeted at managing transient variability and mismatching demand and supply of thermal energy, judicious integration, monitoring and control will become key to the longevity and performance of a TES system in the bigger picture.

Table 1.1. Comparison between the different forms of TES systems [13,16].

	Sensible TES	Latent TES	Thermochemical TES
Storage principle	Sensible heat capacitance of storage media across operating temperature range	Latent heat capacitance arising from phase change of storage medium (PCM) within operating temperature range	Reversible reaction kinetics facilitating energy absorption and release during chemical bond breaking and forming
Relative volumetric energy density	Lowest ($\sim$50 kWh/m^3)	Moderate ($\sim$100 kWh/m^3)	Highest ($\sim$500 kWh/m^3)
Feasible storage time	Short periods; limited by thermal losses to surroundings	Short periods; limited by thermal losses to surroundings	Long periods; unconstrained by surrounding thermal conditions
Thermal energy distribution	Only short distances	Only short distances	Long distances possible
Cost ($/kWh)	0.1–10	10–50	8–100
Technological maturity	Industrial scale	Pilot — industrial scale	Laboratory-scale
Key benefits	• Simple to implement • Low-cost materials • Reliable and well understood	• Higher energy density • Easily implementable with higher customisability than sensible systems	• Maximum possible energy density • Long-term (theoretically indefinite) storage time possible • Storage at ambient conditions; no need for thermal insulation
Key disadvantages	• Low energy density • Short storage duration	• Risk of PCM incompatibilities due to corrosion, thermal expansion during phase change • Short storage duration • Nonlinear scaling of performance with operation	• High complexity • Expensive to implement • Not well understood yet, technology still in its infancy

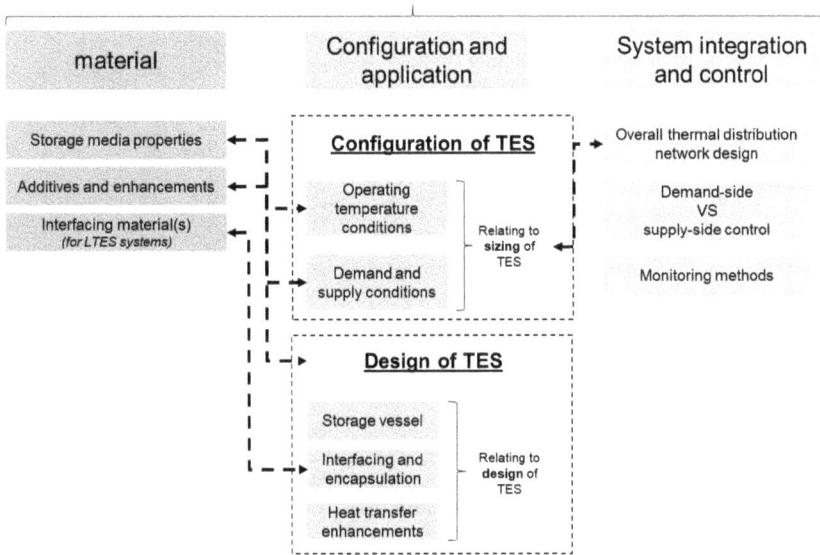

Figure 1.16.   Classifications of the key main foci of TES study.

All aspects of TES study drive at several core criteria for what is considered "favourable" TES behaviour:

(1) All aspects of TES study converge to several core criteria for what is considered favourable TES behaviour;
(2) Maximum possible thermal storage density given the end-user constraints;
(3) Thermally efficient in storage with minimum losses over time;
(4) Cost-efficient at all levels of expense;
(5) Minimum complexity in implementation, monitoring, and control;
(6) High stability in both quantity and quality of discharged thermal energy; and
(7) Results in overall savings in both operational cost and energy savings over time.

The ensuing section will delve briefly into the multiple aspects of TES study that have become key foci of the research community's endeavours. Trends and developments will be briefly discussed to provide a broader perspective on the direction that TES research is headed.

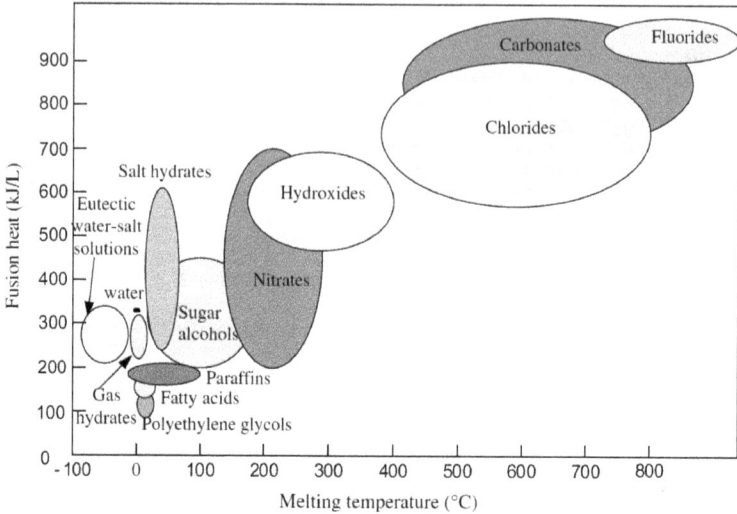

Figure 1.17.   Latent heat across range of PCM types for their respective phase-change temperatures [17].

1.3.1.   *Material aspects of TES study*

Due to the relative simplicity of STES systems, a large portion of TES study is geared towards LTES and TTES systems. PCMs for LTES systems, in particular, have been the subject of much research over time. Figure 1.10 shows the various classes of solid-liquid PCMs that are commonly used in LTES systems. In LTES system design, the phase-change temperature of the selected PCM has to first fall within the operating temperature range of the system before further considerations about its thermophysical properties can be made. Figure 1.17 provides an approximate trend of PCM latent heat to phase-change temperature across all classes of PCMs [17].

Due to the limited availability of high-thermal energy density PCMs for certain temperature ranges, research and industry face considerable difficulty in justifying the use of LTES systems for applications where cost savings from the LTES may not be adequately competitive against that of a STES system of the same size.

Another aspect of material research that is prevalent in the field of PCMs is the use of additives to enhance PCM thermal conductivity and phase-change rate. The phase-change process presents two inherent obstacles: (1) Poor thermal transfer rate throughout the phase transition due to the inherent low thermal conductivity of

PCMs; and (2) supercooling during the solidification process, induced when the PCM is subjected to a high cooling rate. Researchers have devoted substantial effort to resolve the first problem. Novel methods range from packed bed systems to micro-encapsulation have been explored to improve the thermal transfer of PCMs. The second issue is addressed through judicious material processing. Supercooling — the phenomenon where a material subject to a high cooling rate solidifies below its freezing point due to delayed nucleation, has yet to be properly understood by the scientific community [18]. Existing as a meta-stable state, supercooling in LTES systems disrupts thermal performance and creates uncertainties in the monitoring and control of the storage system. Thonon *et al.* [19] describe the supercooling process in 3 stages as shown in Figure 1.18:

(1) Metastable state — as the PCM in the liquid phase is cooled to below the onset of its solidification temperature range, it remains in the liquid phase and continues to only release sensible heat.
(2) Recalescense phase — once the liquid reaches a certain temperature point known as the recalescence starting temperature, the metastable state is broken by heterogenous nucleation when the first crystals appear either at an impurity site or at the interface with the PCM container. Crystallisation then commences, leading to a sharp release of latent heat within the PCM and a resulting increase in the PCM temperature. The temperature

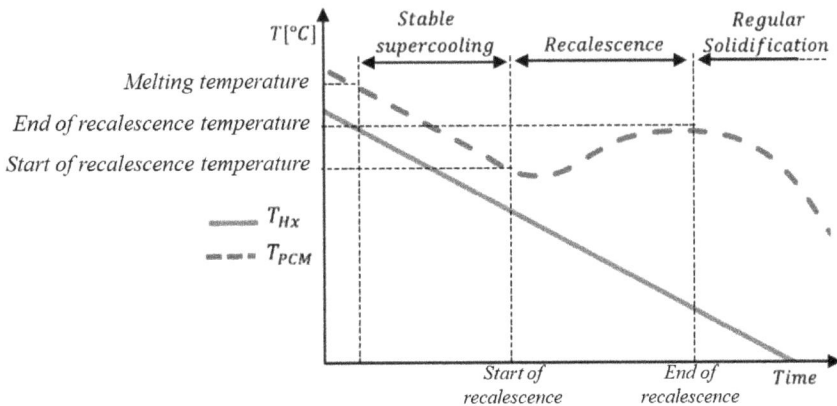

Figure 1.18. The supercooling phenomenon as described by Thonon *et al.* [19]. T_{hx} and T_{pcm} refer to the heat exchange temperature subjected to the PCM and the PCM temperature, respectively.

point at which the recalescence begins is often termed the super-cooling degree.

(3) Regular solidification — once the recalescence phase has ended, the remaining amount of liquid PCM is equivalent to the remaining amount of latent heat that can be extracted from the PCM. Regular solidification proceeds as normal.

Many methods have been proposed to address this phenomenon in LTES systems, of which the use of nanomaterial additives as nucleating agents has become a common approach, despite being met with limited success. Common nanomaterials additives include multi-walled carbon nanotubes (MWCNT), graphene, graphite, and metal or metal-oxide nanoparticles. In addition to being the nucleating sites for PCM solidification, nanoparticle additives also carry the added advantage of enhancing the thermal conductivity of the PCM, showing up to 30% for 1 wt% [20] and up to 65% with 7 wt% [21] of MWCNT in palmitic acid in recent literature (Figure 1.19).

One other aspect of material research which is also specific to LTES systems is the study of "shell" or encapsulation material for encapsulated PCM LTES systems. It has been reported that about 50 different known polymers have been reportedly studied for PCM encapsulation [23], with little attention given to other material types such as inorganics and metals, as organic materials tend to have good chemical and physical stability. Commonly accepted criteria for a good encapsulation material, as reported by Cárdenas-Ramírez,

(a)   (b)

Figure 1.19.   Scanning Electron Microscope imaging of 1% MWCNT (a) and 1 wt% graphene (b) dispersed in eutectic carbonate CPCM [22].

Jaramillo, and Gomez [24] are as follows:

(1) Good mechanical strength
(2) Flexible
(3) Corrosion resistant
(4) Thermal stability
(5) Structural stability and easy handling
(6) Chemical compatibility and non-reactive with the PCM
(7) Non-hygroscopic
(8) Non-toxic
(9) Good availability.

The study of encapsulation materials further extends to the fields of microencapsulation and nanoencapsulation, where sophisticated chemical methods are employed to contain PCM at a micro/nanoscale for improved heat transfer and packing densities. Encapsulation techniques at a smaller scale necessitate more complex encapsulation methods such as chemical methods (polymerisation techniques) and physicochemical methods (sol–gel method, ionic gelation, coacervation), which in turn lead to more complex material property criteria. While this presents a prominent front of academic research in the long term, this text will centre its focus on conventionally applied macroencapsulation methods instead. Several microencapsulation techniques have been classified and illustrated in Figure 1.20 [23].

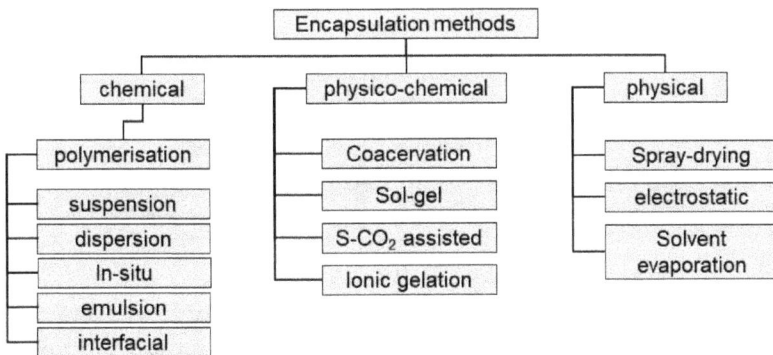

Figure 1.20.   Microencapsulation techniques as classified by Cárdenas-Ramírez *et al.* [24].

1.3.2. *Configuration and design of TES systems*

While material research is a central point of study in thermal engineering fields, the emphasis on efficient sizing and design of the overall TES is often the subject of much debate. For this section, the
configuration of the TES system refers to the method of assessing its
requisite capacity and storage size for the application, while design
refers specifically to the study of its internal structure and interfacing features. STES systems such as hot water or chilled water
storage tanks are utilised extensively in district and building thermal systems where several constraints and variability in operation
exist. Given the lack of variables and parameters for its operation
and design, the focus of many studies on hot/chilled water storage
tanks is centred on diffuser design and stratification management,
as the efficiency of these TES systems relies on minimising mixing
within the fluid media during charging and discharging processes.
Configurations of TES tanks for specific heating and cooling applications can vary depending on the scale of demand or supply. For
instance, Chandra and Matsuka [25] define three common configurations of hot water storage tanks for domestic utility, as shown in
Figure 1.21.

Directly heated storage tanks like that featured in Figure 1.21(a)
tend to not only possess the highest thermal transfer rates but also
the highest degrees of fluid mixing and turbulence, resulting in better
management of stratification with the incorporation of baffle plates
and diffusers. Van Ruth [26] presented a thermo-differential self-
actuating valve to adjust the flow into the storage tank depending

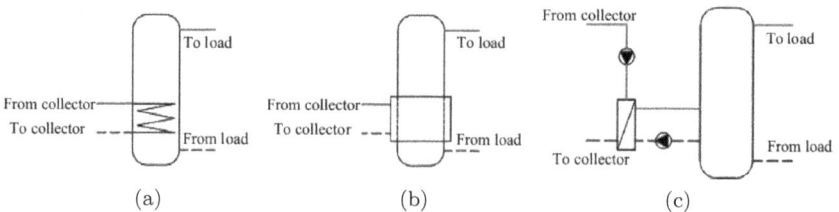

Figure 1.21.   Configurations of domestic hot water (DHW) TES tanks such as
(a) immersed heat exchanger configuration, (b) mantle heat exchanger configuration, and (c) external heat exchanger configuration [25].

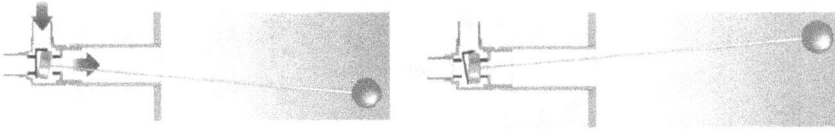

Figure 1.22. Thermo-differential inlet valve as designed and investigated by Ruth [26].

on the tank internal conditions as shown in Figure 1.22. Wang and Davidson [27] studied a porous-tube thermal stratifier manifold for a solar DHW TES system. Dragsted *et al.* [28] experimentally investigated multiple inlet stratifiers to understand their relative impacts on stratification in water storage tanks. On the other hand, indirect heating configurations presented in Figures 1.21(b) and 1.21(c) tend to display poorer thermal transfer due to the additional heat transfer loop but are able to manage stratification well [25].

LTES systems, especially encapsulated PCM LTES systems add a separate dimension of design to the fray. Appropriate encapsulation sizing, optimisation and selection is a crucial and much-studied part of the encapsulated PCM LTES design. As shown in Figure 1.12, PCM encapsulation can assume many forms such as tubes, slabs, and spheres. Spherical encapsulation packed-bed systems, in particular, have been widely studied in the literature due to their potential to yield high energy density and surface-area-to-volume ratio, as well as ease of installation and maintenance. For slab-type encapsulation methods, however, more organised methods of deployment are required as space optimization in vertical cylindrical storage tanks can prove to be challenging, as highlighted in Figure 1.23.

Many shell-and-tube setups involving the tube-side HTF and shell-side PCM [29–33], as displayed in Figure 1.24, have been widely studied. However, the encapsulation form has been relatively unexplored besides characterisation studies involving single tubes–HTF setups [34]. Zauner *et al.* [35] correctly pointed out that studies on tube-encapsulated PCM have been few and thus conducted extensive experimental testing on a full-scale high-temperature LTES system involving paraffin-based PCM contained within the tubes of a baffled shell-tube heat exchanger operating as a storage tank as shown in Figure 1.25. In other works, Xiao and Zhang [36,37] studied the

Figure 1.23. Commercial deployment of slab-type PCM encapsulations [11]. Due to the aspect ratio of slab-type encapsulations, horizontal or non-cylindrical storage vessels tend to be used instead of conventional vertical cylindrical storage tanks.

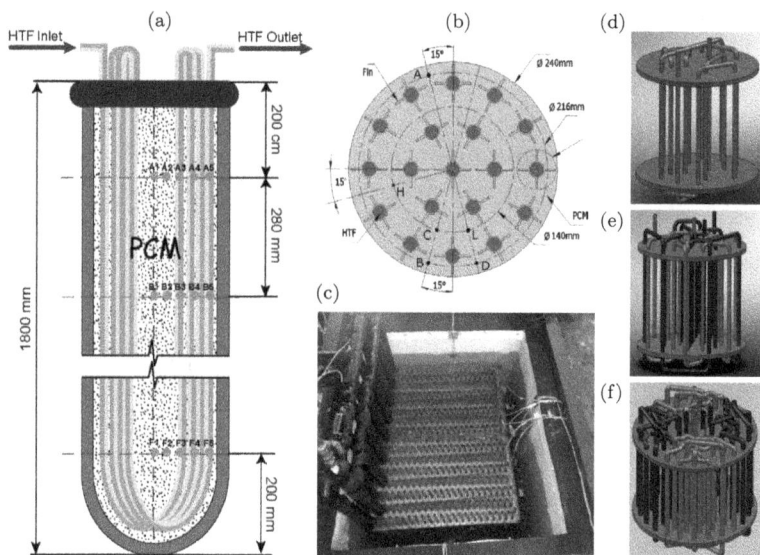

Figure 1.24. Studies on shell-and-tube EPCM LTES systems involving the PCM on the shell side and HTF on the tube side [30–33]. (a) U-tube shell-and-tube storage unit as designed by Osman *et al.* [30]; (b) ice-on-coil storage as experimented by Abhishek *et al.* [32]; (c) cross-section of shell-and-tube unit studied by Abreha *et al.* [31]; (d) single-tube; (e) double-tube; and (f) quadruple-tube configurations as tested by Tay *et al.* [33].

charging and discharging performance of tube-encased PCM in a vertical cylindrical LTES without baffles. Tao and He [38] compared the performances of tube-side and shell-side PCM in a singular unit and found that tube-side PCM configurations allowed for higher heat transfer rates without compromising overall storage capacity even in the absence of natural convection.

Figure 1.25. Shell-tube exchanger configuration (with baffles) LTES with PCM encased on the tube side as studied by Zauner *et al.* [35]. To the author's knowledge, this has been the only extensive study conducted on this format of PCM encapsulation to date.

1.3.3. *Application and system integration of TES systems*

TES systems are applied to a wide range of situations. Depending on the user's requirements, it can be tricky to optimise TES sizing and operation for the specific role it plays in a thermal distribution system. One principal focus of this book is on the application of TES in district thermal management systems. Air conditioning and mechanical ventilation (ACMV) occupy an outsized proportion of the energy demand in urban landscapes, especially in tropical climates where ACMV energy demand itself accounts for more than 50% of urban energy demand. As a result, TES systems are employed to reduce chiller operational expenses by charging TES systems during off-peak electrical tariffs, while allowing the TES systems to support most of the load during discharging during peak-hour tariffs. TES can be integrated into district cooling systems (DCS) in a multitude of ways. Gang *et al.* [39] listed three most common ways that TES systems are incorporated into DCS as shown in Figure 1.26. In the upstream chiller configuration, the user load is supported by either the chiller or TES directly or both at the same time depending on the desired operating scheme. In the downstream chiller configuration,

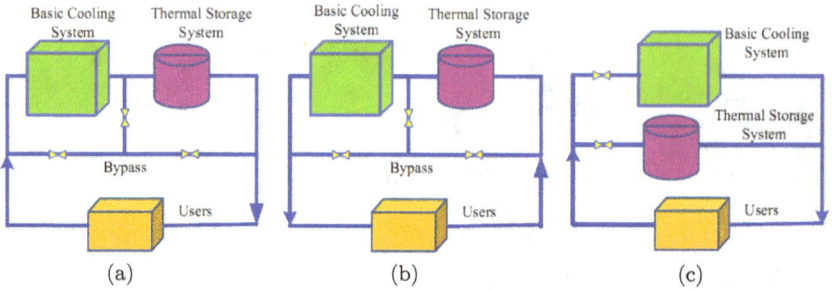

Figure 1.26.   (a) Serial connection with chillers upstream; (b) serial connection with chillers downstream; and (c) parallel connection [39].

the TES is employed to reduce the cooling load that would otherwise be supported by the chiller by reducing the return temperature entering the chiller. As the chiller's part-load efficiency is often lower than its full-load efficiency in most cases, the additional capital incurred in this scenario is largely offset by the initial capital cost savings on a smaller-capacity chiller. In the parallel connection system, both operation modes are possible but elaborate piping and pumping might be necessary to preserve stratification in the tank when the operation modes are changed.

Additionally, the sizing of the TES relative to the capacity of the DCS or associated cooling/refrigeration system must also be decided in various ways depending on the desired operating strategy. Selvnes *et al.* [40] and Yau *et al.* [41] have described this with three distinct operation strategies for TES in refrigeration systems as shown in Figure 1.27.

Strategy (b) involving partial storage with load levelling is often considered the ideal case in TES sizing, especially in brownfield installations as it allows the chiller loading to remain relatively constant throughout operation. This will reduce inefficiencies arising from part-load situations which also realizes a lower chiller capacity and hence lower initial capital investment as compared to strategies (a) and (c). Aside from cooling applications, TES systems are also relevant in heating systems where there is also a possibility of incorporating renewable processes, such as solar-thermal energy generation shown in Figure 1.28 [42].

While operating strategies statically dictate the intended TES operating configuration, active control systems actually determine

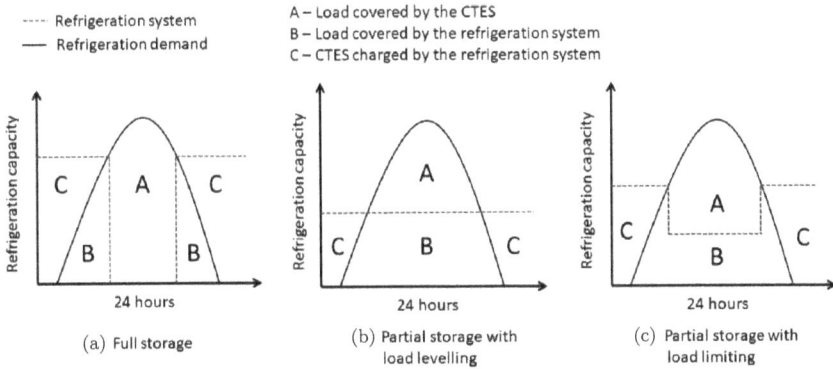

Figure 1.27. (a) Full storage — the TES completely supports cooling load during peak periods, (b) partial storage with load levelling — the TES supports part of the cooling load during peak hours, allowing the chiller to maintain constant load throughout operation period, (c) partial storage with load limiting — the TES supports part of the cooling load during peak hours but is sized to prevent the chiller load from exceeding a specific limit [40].

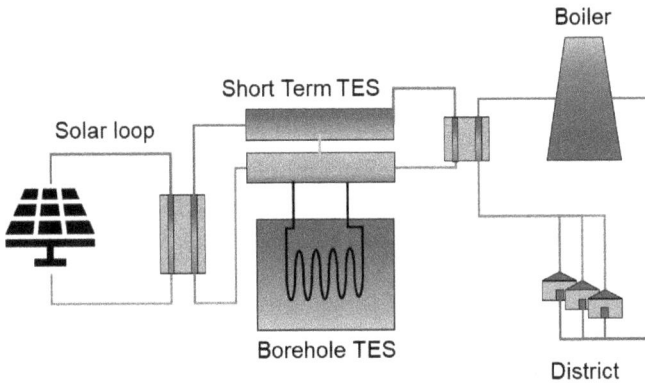

Figure 1.28. TES systems integrated with a district heating system involving solar thermal collectors [42].

the TES operating manner within its system and the varying external conditions it is subjected to. Sun *et al.* [43] classified the load-shifting control of cooling systems as either being heuristic or optimal control as depicted in Figure 1.29. It was further concluded that simple control strategies underperformed when they are compared to optimal control strategies even under simple tariff pricing.

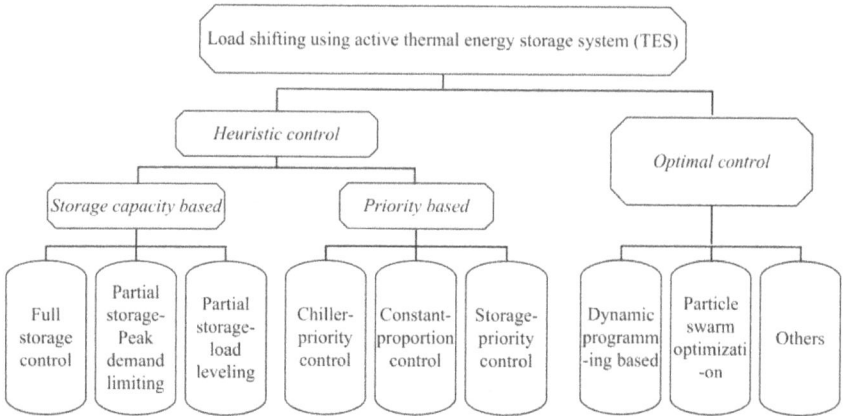

Figure 1.29.   Classification of load shifting control in cooling systems with TES by Sun *et al.* [43].

Cox *et al.* [44] similarly noted that the popularity of using conventional static control rules has been surpassed by optimal control strategies that involve machine learning for load prediction and metaheuristic optimization. Machine learning methods such as neural networks (NN) have gained popularity largely due to their ease of use, ability to generalise data and the wide availability of software support. Metaheuristic optimization methods include genetic algorithms and swarm-intelligence algorithms, which are often used to address unconstrained optimization problems. Cox *et al.* [44] implemented a demand-response model-predictive control (MPC) scheme comprising a nonlinear autoregressive with exogenous output NN for load forecasting, a physics-based model of the cooling system, and a genetic algorithm optimization strategy for controlling a university campus-wide cooling system. The MPC scheme was demonstrated to generate significant cost savings of up to 15.9–16.4% for both time-of-use and real-time pricing tariffs (Figure 1.30).

Besides thermal distribution networks, one application that has been gaining traction is the use of TES for the regasification of liquefied natural gas (LNG) in power plants. LNG for power plants is stored under cryogenic conditions of $-163°C$ at atmospheric pressure. Before combustion, the LNG is first pressurised to 5 bar and then regasified to ambient temperature through LNG vaporisers. Common LNG vaporisers include open-rack vaporisers (ORV)

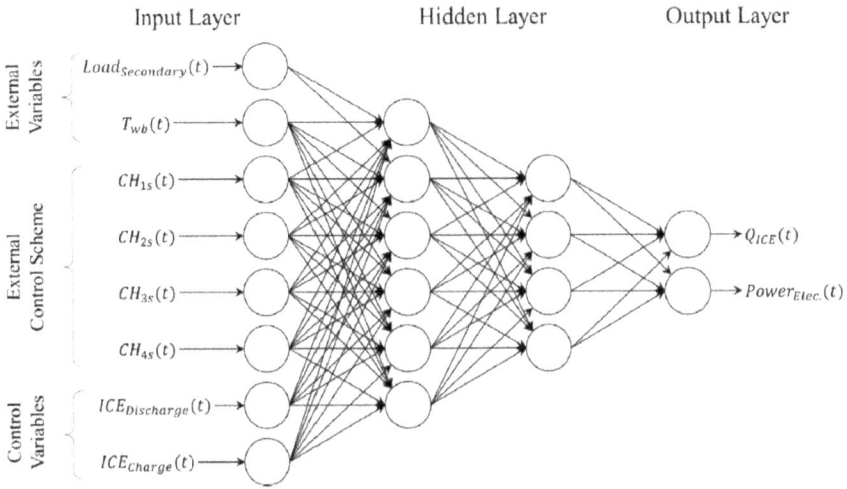

Figure 1.30. Neural network used to estimate the chiller performance as implemented by Cox *et al.* [44].

or submerged-combustion vaporisers (SCV), which employ seawater or waste heat as the heat source. As a result, a large quantity of high-grade "cold" energy is often lost to ambient conditions. In recent times, a significant amount of research effort has been devoted to addressing this by developing technologies for "cold" energy recovery from LNG regasification. Khor *et al.* [45] estimated up to 860 kJ/kg of cold energy or 370 kJ/kg of cold exergy is recoverable from the regasification process. To date, a large portion of the focus on cryogenic "cold" energy recovery from LNG regasification has centred on condensation for closed-loop low-temperature power cycles such as Brayton, Kalina, and Organic Rankine Cycles (ORC). Due to the low operating temperature range, working fluids like propane, noble gases, and refrigerants are often used. Other attempts at "cold" energy recovery involve various forms of gas liquefaction, sublimation, or separation. In comparison, the use of direct TES systems for LNG regasification is only in its infancy, as fundamental problems concerning this approach have yet to be addressed. Conventional end-user temperature range of most cooling-application TES spans $-4°C$ to $12°C$, compared to the entire regasification ranging $-163°C$ to $25°C$. Sizing the TES to operate closer to the end-user range would lead to extremely high losses in cold exergy during LNG regasification, as the quality of cold is lost. On the contrary, storage at temperatures closer to cryogenic conditions retains more of the cold exergy but results in inadequate regasification. One way to address this is by designing multi-stage cold extraction systems, where the exergy losses can be minimised by reducing the temperature gain across each stage. However, this approach is capital intensive due to the infrastructure requirements for thermal transport and insulation, especially since storage at lower temperatures can incur huge losses to the ambient over protracted periods of time. Khor *et al.* [45] estimated a 4-stage cold recovery process using LTES systems with PCMs over 4 separate phase-change temperatures ($-86°C$, $-67°C$, $-33°C$, and $-3°C$) to yield a maximum exergy efficiency of 24% with over 75% of the cold exergy being destroyed in the process (Figure 1.31). A more recent work by Shao *et al.* [46] proposed a multi-objective genetic algorithm to optimize a multi-stage

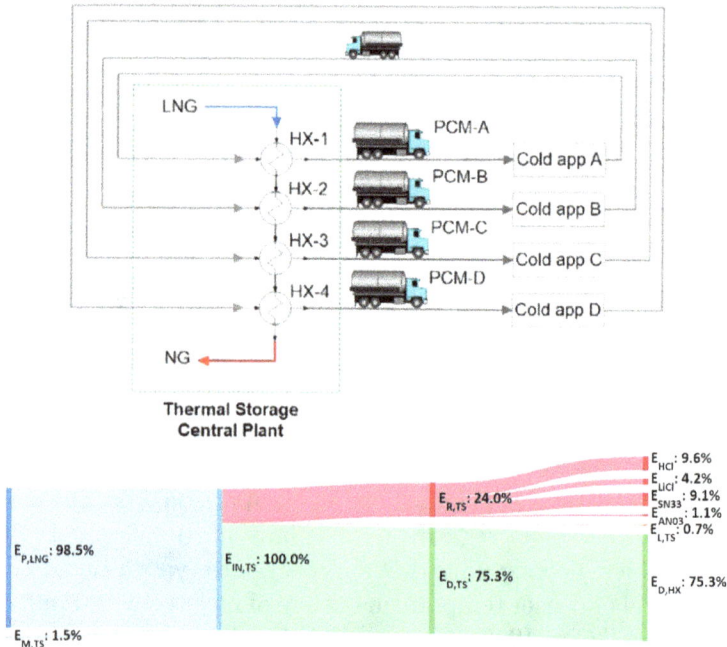

Figure 1.31.   Exergy flows for a 4-stage PCM LTES system for LNG regasification "cold" recovery by Khor *et al.* [45]. Up to 75% of the cold exergy from the LNG was destroyed even with a 4-stage process.

LNG regasification process, as shown in Figure 1.32, involving an LTES storage process, a subcritical CO_2 ORC process, a chilled water cooling and air dehumidification stage. Key results from this work revealed that the LTES exergy efficiency of up to 21.7% was possible if the system design was optimized solely for exergy efficiency maximisation.

Tafone *et al.* [47] noted that works related to both numerical and experimental studies for low-temperature cryogenic TES systems are few. Many existing studies for high-grade cold storage for applications such as liquid air energy storage (LAES) involve sensible TES systems due to their simplicity and low cost. A study has been conducted on a packed bed sphere-encapsulated PCM LTES for LAES cold energy storage to assess its suitability for the application. The study concluded that a payback period of 3 years was possible when the local electrical off-peak tariff was designated at $0.09/kWh$_e$ (Figure 1.33).

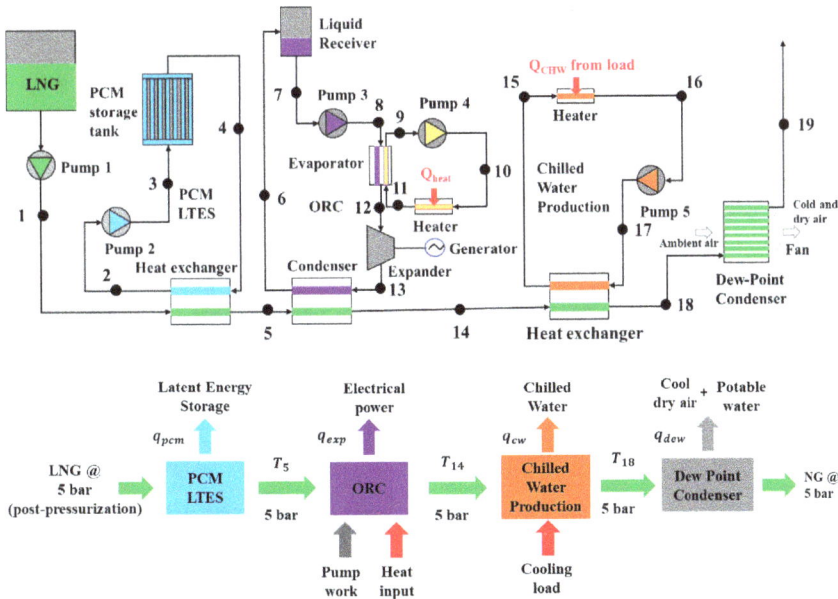

Figure 1.32.   Multi-stage LNG regasification process with LTES, ORC, chilled water production, and air dehumidification by Shao *et al.* [46].

Figure 1.33.   Experimental setup by Tafone *et al.* [47].

1.4.   Conclusion

In this chapter, a general overview of TES systems, their importance to the thermal systems design, and their rising importance was discussed. A brief breakdown of their various manifestations and embodiments was also conducted with an emphasis on recent developments. The core thrusts of TES study, namely, material, design, and system integration, were defined and their research directions were briefly reviewed. The subsequent chapters will provide greater details on different forms of TES, starting from the fundamentals to present-day case studies.

As the energy landscape continues to evolve with higher levels of variability from climate change and renewable energy sources, the progress of TES systems is expected to grow significantly including various approaches to better manage thermal systems. While heat storage was central to the thermal management of the past systems, rising global temperatures have led to the growing need for cold TES systems in places where ambient temperatures continue to appreciate. Accordingly, a certain degree of emphasis will be placed on cold-TES systems in a subsequent chapter.

References

[1] Centre Tecnològic de Transferència de Calor (CTTC), Thermal Energy Storage for CSP plants, Centre Tecnològic de Transferència de Calor (CTTC) [Online]. Available: https://www.cttc.upc.edu/?q=research/node/94.

[2] Dhruva Technologies Pvt. Ltd, PCM based Thermal Energy Storage (TES), Dhruva Technologies Pvt. Ltd 2022 [Online]. Available: http://dhruvatec.com/phase-change-material.html.

[3] A. Wilson, Buildings on Ice: Making the case for thermal energy storage, BuildingGreen, Inc., 30 June 2009 [Online]. Available: https://www.buildinggreen.com/feature/buildings-ice-making-case-thermal-energy-storage.

[4] E. Guelpa and V. Verda, Thermal energy storage in district heating and cooling systems: A review, *Applied Energy*, vol. 252, p. 113474, 2019.

[5] IRENA, Innovation Outlook: Thermal Energy Storage, International Renewable Energy Agency, Abu Dhabi, 2020.

[6] Keppel Corp, Overview, Keppel Corp, 2010 [Online]. Available: https://www.keppeldhcs.com/our_product_overview.html.

[7] B. Koçak, A. I. Fernandez and H. Paksoy, Review on sensible thermal energy storage for industrial solar applications and sustainability aspects, *Solar Energy*, vol. 209, pp. 135–169, 2020.

[8] S. Agarwala and K. N. Prabhu, Review of thermal characterization techniques for salt-based phase change materials, *Journal of Energy Storage*, vol. 46, p. 103865, 2022.

[9] A. Fallahi, G. Guldentrops, M. Tao, S. Granados-Focil and S. Van Dessel, Review on solid-solid phase change materials for thermal energy storage: Molecular structure and thermal properties, *Applied Thermal Engineering*, vol. 127, pp. 1427–1441, 2017.

[10] J. Lei, J. Yang and E.-H. Yang, Energy performance of building envelopes integrated with phase change materials for cooling load reduction in tropical Singapore, *Applied Energy*, vol. 162, pp. 207–217, 2016.

[11] PCM Products Ltd, Our standard range of Encapsulated PCM Products, PCM Products Ltd [Online]. Available: https://www.pcmproducts.net/Encapsulated_PCMs.htm.

[12] Rubitherm Technologies GmbH, Macroencapsulation — CSM, Rubitherm Technologies GmbH [Online]. Available: https://www.rubitherm.eu/en/index.php/productcategory/makroverkaspelung-csm.

[13] F. Desai, S. Prasad Jenne, P. Muthukumar and M. M. Rahman, Thermochemical energy storage system for cooling and process heating applications: A review, *Energy Conversion and Management*, vol. 229, p. 113617, 2021.

[14] Fraunhofer Institute for Solar Energy Systems ISE, Ecological Evaluation of Selected Storage Materials and Concepts for Heating and Cooling Applications (Speicher-LCA), Fraunhofer Institute for Solar Energy Systems ISE, 2019 [Online]. Available: https://www.ise.fraunhofer.de/en/research-projects/speicher-lca.html.

[15] J. Stengler, I. Bürger and M. Linder, Performance analysis of a gas-solid thermochemical energy storage using numerical and experimental methods, *International Journal of Heat and Mass Transfer*, vol. 167, p. 120797, 2021.

[16] A. Safari, R. Raidur, F. Sulaima, Y. Xu and J. Dong, A review on supercooling of phase change materials in thermal energy storage systems, *Renewable and Sustainable Energy Reviews*, vol. 70, pp. 905–919, 2017.

[17] G. Li, Y. Hwang, R. Radermacher and H.-H. Chun, Review of cold storage materials for subzero applications, *Energy*, vol. 51, pp. 1–17, 2013.

[18] D. Lilley, J. Lau, C. Dames, S. Kaur and R. Prasher, Impact of size and thermal gradient on supercooling of phase change materials for thermal energy storage, *Applied Energy*, vol. 290, p. 116635, 2021.

[19] M. Thonon, G. Fraisse, L. Zalewski and M. Pailha, Analytical modelling of PCM supercooling including recalescence for complete and partial heating/cooling cycles, *Applied Thermal Engineering*, vol. 190, p. 116751, 2021.

[20] J. Wang, H. Xie, Z. Xin, Y. Li and L. Chen, Enhancing thermal conductivity of palmitic acid based phase change materials with carbon nanotubes as fillers, *Solar Energy*, vol. 84, pp. 339–344, 2010.

[21] P. Ji, H. Sun, Y. Zhong and W. Feng, Improvement of the thermal conductivity of a phase change material by the functionalized carbon nanotubes, *Chemical Engineering Science*, vol. 81, pp. 140–145, 2012.

[22] Y. Tao, C. Lin and Y. He, Preparation and thermal properties characterization of carbonate salt/carbon nanomaterial composite phase change material, *Energy Conversion and Management*, vol. 97, pp. 103–110, 2015.

[23] J. Giro-Paloma, M. Martinez, L. F. Cabeza and A. Ines Fernandez, Types, methods, techniques, and applications for microencapsulated phase change materials (MPCM): A review, *Renewable and Sustainable Energy Reviews*, vol. 53, pp. 1059–1075, 2016.

[24] C. Cárdenas-Ramírez, F. Jaramillo and M. Gomez, Systematic review of encapsulation and shape-stabilization of phase change materials, *Journal of Energy Storage*, vol. 30, p. 101495, 2020.

[25] Y. P. Chandra and T. Matsuka, Stratification analysis of domestic hot water storage tanks: A comprehensive review, *Energy & Buildings*, vol. 187, pp. 110–131, 2019.

[26] N. J. van Ruth, New type of valve for solar thermal storage tank stratification, *Energy Procedia*, vol. 91, pp. 246–249, 2016.

[27] S. Wang and J. H. Davidson, Selection of permeability for optimum performance of a porous tube thermal stratification manifold, *Solar Energy*, vol. 122, pp. 472–485, 2015.

[28] J. Dragstead, S. Furbo, M. Dannemand and F. Bava, Thermal stratification built up in hot water tank with different inlet stratifiers, *Solar Energy*, vol. 147, pp. 414–425, 2017.

[29] G. Shen, X. Wang and A. Chan, Experimental investigation of heat transfer characteristics in a vertical multi-tube latent heat thermal energy storage system, *Energy Procedia*, vol. 160, pp. 332–339, 2019.

[30] M. Osman, M. Abokersh, O. El-Baz, O. Sharaf, N. Mahmoud and M. El-Morsi, Key performance indicators (KPIs): Assessing the process integration of a shell-and-tube latent heat storage unit, *Journal of Cleaner Production*, vol. 256, p. 120249, 2020.

[31] B. G. Abreha, P. Mahanta and G. Trivedi, Thermal performance evaluation of multi-tube cylindrical LHS system, *Applied Thermal Engineering*, vol. 179, p. 115743, 2020.

[32] A. Abhishek, B. Kumar, M. H. Kim, Y. T. Lee, J. D. Chung, T. S. Kim, T. Kim, C. Lee and K. Lee, Comparison of the performance of ice-on-coil LTES tanks with horizontal and vertical tubes, *Energy & Buildings*, vol. 183, pp. 45–53, 2019.

[33] N. Tay, M. Belusko and F. Bruno, An effectiveness-NTU technique for characterising tube-in-tank phase change thermal energy storage systems, *Applied Energy*, vol. 91, pp. 309–319, 2012.

[34] H. Liang, J. Niu and Y. Gan, Performance optimization for shell-and-tube PCM thermal energy storage, *Journal of Energy Storage*, vol. 30, p. 101421, 2020.

[35] C. Zauner, F. Hengstberger, B. Mörzinger, R. Hofmann and H. Walter, Experimental characterization and simulation of a hybrid sensible-latent heat storage, *Applied Energy*, vol. 189, pp. 506–519, 2017.

[36] X. Xiao and P. Zhang, Numerical and experimental study of heat transfer characteristics of a shell-tube latent heat storage system: Part I — Charging process, *Energy*, vol. 79, pp. 337–350, 2015.

[37] X. Xiao and P. Zhang, Numerical and experimental study of heat transfer characteristics of a shell-tube latent heat storage system: Part II — Discharging process, *Energy*, vol. 80, pp. 177–189, 2015.

[38] Y. Tao, Y. Liu and Y.-L. He, Effects of PCM arrangement and natural convection on charging and discharging performance of shell-and-tube LHS unit, *International Journal of Heat and Mass Transfer*, vol. 115, pp. 99–107, 2017.

[39] W. Gang, S. Wang, F. Xiao and D.-C. Gao, District cooling systems: Technology integration, system optimization, challenges and opportunities for applications, *Renewable and Sustainable Energy Reviews*, vol. 53, pp. 253–264, 2016.

[40] H. Selvnes, Y. Allouche, R. I. Manescu and A. Hafner, Review on cold thermal energy storage applied to refrigeration systems using phase

change materials, *Thermal Science and Engineering Progress*, vol. 22, p. 100807, 2021.

[41] Y. Yau and B. Rismanchi, A review on cool thermal storage technologies and operating strategies, *Renewable and Sustainable Energy Reviews*, vol. 16, pp. 787–797, 2012.

[42] H. Quintana and M. Kummert, Optimized control strategies for solar district heating, *Journal of Building Performance Simulation*, vol. 8, pp. 79–96, 2015.

[43] Y. Sun, S. Wang, F. Xiao and D. Gao, Peak load shifting control using different cold thermal energy storage facilities in commercial buildings: A review, *Energy Conversion and Management*, vol. 71, pp. 101–114, 2013.

[44] S. J. Cox, D. Kim, H. Cho and P. Mago, Real time optimal control of district cooling system with thermal energy storage using neural networks, *Applied Energy*, vol. 238, pp. 466–480, 2019.

[45] J. Khor, F. Dal Magro, T. Gundersen, J. Sze and A. Romagnoli, Recovery of cold energy from liquefied natural gas regasification: Applications beyond power cycles, *Energy Conversion and Management*, vols. 336–355, p. 174, 2018.

[46] Y. Shao, K. Soh, Y. Wan, Z. Huang, M. Islam and K. Chua, Multiobjective optimization of a cryogenic cold energy recovery system for LNG regasification, *Energy Conversion and Management*, vol. 244, p. 114524, 2021.

[47] A. Tafone, E. Borri, L. F. Cabeza and A. Romagnoli, Innovative cryogenic Phase Change Material (PCM) based cold thermal energy storage for Liquid Air Energy Storage (LAES) — Numerical dynamic modelling and experimental study of a packed bed unit, *Applied Energy*, vol. 301, p. 117417, 2021.

Chapter 2

Sensible Thermal Energy Storage Systems

Abstract

Sensible thermal energy storage (STES) systems constitute the most established and widely applied form of thermal energy storage in industry. In this chapter, the fundamentals of STES systems as well as the various aspects of their performance analysis will be first covered. Following which, case studies will be delved into and recent innovations and studies involving both commercial and academic fields will be explored.

Keywords: Sensible thermal energy storage systems; thermal stratification; numerical methods; diffusers.

2.1. Fundamentals of Sensible Thermal Energy Storage (STES) Systems

In STES systems, the specific heat capacity of the storage media is used to store heat (or cold) within its operating temperature range. A key determinant of the system's storage capacity is the sensible heat capacity of the storage media within the STES. As such, materials of higher specific heat capacity contribute to higher thermal energy densities of the STES. Key properties for consideration in material selection for STES systems include the following:

  i. Density
 ii. Specific heat
iii. Thermal conductivity
 iv. Thermal diffusivity
  v. Chemical compatibility with container/tank material
 vi. Chemical stability.

2.1.1. *Basic thermodynamics of STES systems*

Gonzalez-Roubaud *et al.* [1] evaluated the performance of a STES via the following aspects:

(1) Storage capacity, which is the total thermal energy stored within the system;
(2) Power, which is the rate at which the thermal energy can be stored into or supplied from the STES;
(3) Efficiency, which is defined as the ratio of energy delivered throughout discharge to the total energy charged;
(4) Charge/discharge time, which defines the cycling time between charge and discharge processes of the STES; and
(5) Cost, which refers to the total installation and operation costs of the STES. This is often evaluated relative to the storage capacity of the STES or its total volume, depending on key comparison constraints when studied.

The amount of energy stored in a STES is computed simply by the total sensible heat for all storage media used, within its operating temperature as

$$E = \sum_i m_i C_{p,i} \Delta T_{\text{op}} \tag{2.1}$$

$$\Delta T_{\text{op}} = T_H - T_C \tag{2.2}$$

where E (J) refers to the amount of thermal energy stored, m (kg) refers to the total mass of the respective storage media within the STES, C_p (J/kg $\cdot$ K) refers to the specific heat capacity of the storage media within the tank, and ΔT_{op} (°C) refers to the operating temperature range of the STES defined by Equation (2.2). The temperatures T_H and T_C (°C) refer to the upper and lower bounds of the operating temperature range, respectively, and the subscript i refers to each storage medium used within the STES.

As shown in Figure 2.1, the charging and discharging power for the STES can then be defined as

$$Q_C = \dot{m}_C C_p (T_{\text{out}} - T_{\text{in}}) \tag{2.3a}$$

$$Q_D = \dot{m}_D C_p (T_{\text{out}} - T_{\text{in}}) \tag{2.3b}$$

where $\dot{m}_C$ and $\dot{m}_D$ (kg/s) refer to the respective charging and discharging mass flow rates through the STES. Based on this, the total charge and discharge capacity, as shown in Figure 2.2, is then defined as

$$E_C = \int_0^{t_C} Q_C dt \tag{2.4a}$$

$$E_D = \int_0^{t_D} Q_D dt \tag{2.4b}$$

where t_C and t_D (s) refer to the charging and discharging time, respectively. The resultant efficiency of the STES is then defined as the ratio of discharged to charged capacity, given as

$$\eta_{\text{TES}} = \frac{E_D}{E_C} \tag{2.5}$$

This is sometimes referred to as the figure of merit (FOM). While this metric sufficiently characterizes the "true" efficiency of the TES, it is often difficult to measure in practice as most TES systems do

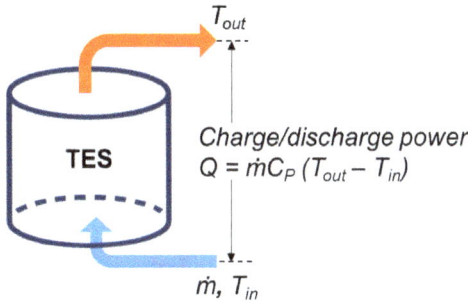

Figure 2.1.   Schematic representation of the charge/discharge power as the net rate of enthalpy across the TES.

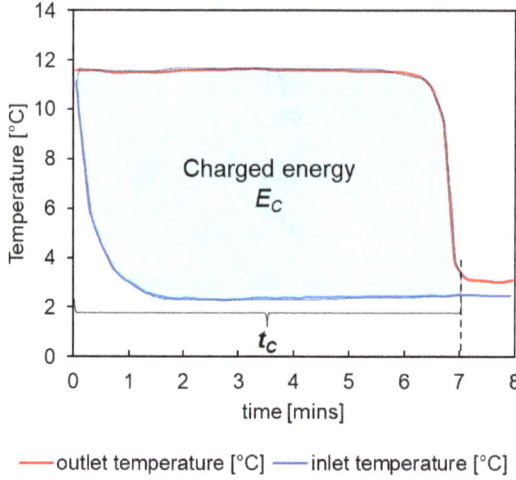

Figure 2.2.   The charged energy in an experimental chilled water TES over the test period as studied by Oro *et al.* [2], highlighted in blue.

not have the capacities for prolonged periods of discharge. A more common approach to execute is via the half-cycle figure of merit (half FOM), which measures the ratio of the useful thermal energy capacity to the theoretical capacity described by Majid *et al.* [3] as

$$\text{FOM}_{\frac{1}{2}} = \frac{E_{\text{int}}}{E} \tag{2.6a}$$

$$E_{\text{int}} = \frac{\rho A_c C_P (T_H - T_C)}{S}(\log(1 + 10^{SC}) - \log 2) \tag{2.6b}$$

where A_C (m^2) refers to the cross-sectional area of the tank and ρ (kg/m^3) refers to the density of the TES storage media. The values of S and C (m) are thermocline position and gradient, respectively, as shown in Figure 2.3. These are derived from the TES internal temperature distribution based on a sigmoid-dose response (SDR) function described as

$$T(y) = T(C) + \frac{T_H - T_C}{1 + 10^{(C-y)S}} \tag{2.7}$$

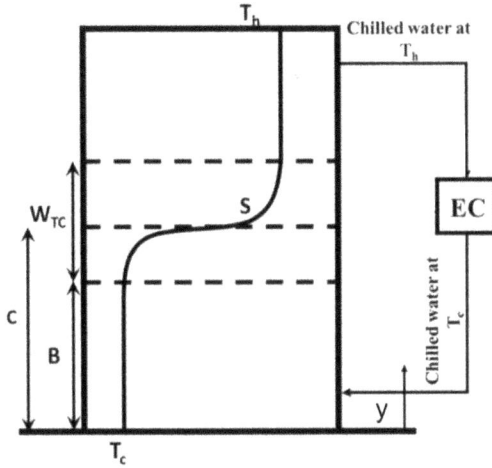

Figure 2.3. Depiction of the thermocline represented by an SDR function and the associated fitting constants C and S [4].

2.1.2. *Stratified STES systems*

The key determinant of the STES efficiency, especially in single-fluid systems like hot and cold water tanks, lies in the stratification within the fluid. As the warmer fluid becomes more separated from the cold fluid at the bottom of the LTES, less heating or cooling utility gets wasted arising from thermal mixing within the mixing region. For maximum efficiency, this degree of thermal stratification should be maintained as much as possible during storage as well as during charge–discharge processes. Different degrees of stratification are as represented in Figure 2.4.

The stratification within STES tends to be influenced by several key design factors, namely, height–width aspect ratio, diffuser design, and basal storage media thermophysical properties.

The height–diameter aspect ratio of the STES impacts thermal stratification by balancing the degree of mixing within the fluid against the thermal losses to the ambient. Small aspect ratios have been known to result in a higher degree of mixing between the warm and cold fluids. In contrast, TES tanks with higher aspect ratios promote stratification at the cost of higher ambient heat losses. Early work by Lavan and Thompson [6] showed that there are marginal gains from TES efficiency when the height–diameter

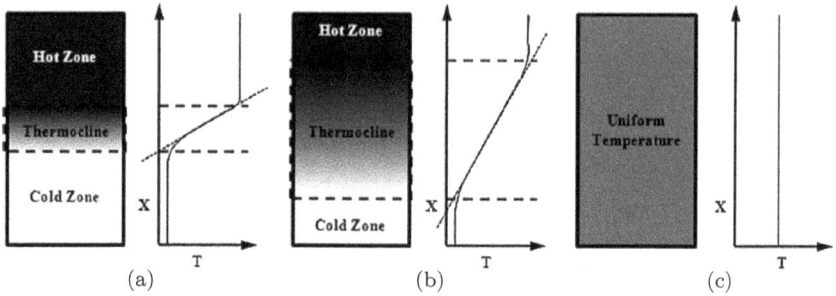

Figure 2.4.   Stratification within STES tanks under (a) highly stratified conditions; (b) weakly stratified conditions; and (c) completely mixed and un-stratified conditions [5].

ratio was increased from 3 to 4, while Castell *et al.* [7] found that an aspect ratio of 3.3 was sufficient for many analytical studies in the literature.

Stratifier design is another well-known aspect of fundamental STES study. When a fluid flow is introduced to the existing bulk fluid within a STES, the cooler fluid tends to sink, while lower-density warmer fluid moves to the higher parts of the tank. This is, however, heavily influenced by the momentum at which the fluid enters the TES. Higher speed and more directed fluid flow promotes mixing, while slower, distributed flow causes less disturbance to the fluid layers. The resulting thermocline region between the temperature layers of the TES is then approximated by the various distribution models such as Stepped, Linear, or Three-Zone models or even complex functions such as the SDR function as shown in Equation (2.7). While the thickness of the thermocline region determines the efficiency of the TES, it also serves as a thermal barrier to separate the hot and cold layers, delaying further mixing in the TES [9]. Types of stratifiers that have been studied in the literature include baffles [10], diffusers [8], fabrics, and membranes [11]. Figures 2.5 and 2.6 show a few of the selected tank stratifier configurations that have been explored in the literature.

Furbo [13], in his study, concluded that the fluid's inherent properties are key to thermal stratification within the TES. Key properties such as density and thermal diffusivity play crucial roles in the stability of the thermocline within the TES. Equations (2.8a)–(2.8e) highlight how these properties are dependent on temperature

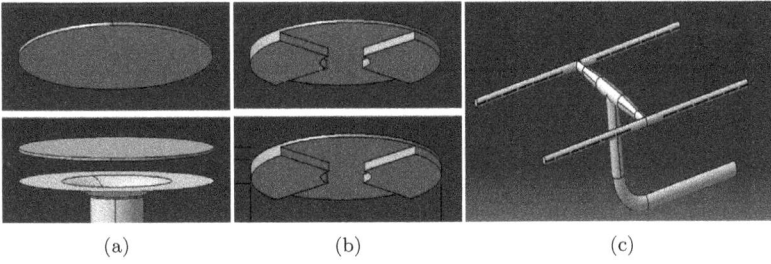

(a)   (b)   (c)

Figure 2.5. Diffuser types studied by Chung *et al.* [8]. (a) Radial plate; (b) radial adjusted plate; and (c) H-beam.

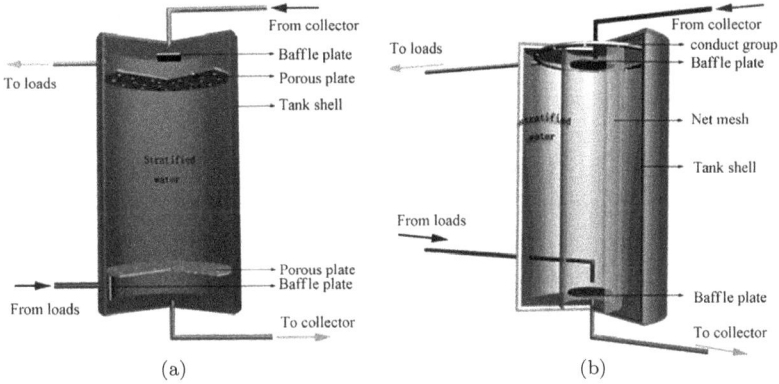

(a)   (b)

Figure 2.6. Structural and design comparison between stratified TES tanks fitted with (a) baffle plate and (b) porous structure by Mangold and Deschaintre [12].

in water.

$$\rho = 1000.6 - 0.0128T^{1.76} \tag{2.8a}$$

$$\nu = 1.477 \times 10^{-6} \exp(-1.747 \cdot 10^{-2}) \tag{2.8b}$$

$$k = 0.52 + 0.0198T^{0.46} \tag{2.8c}$$

$$C_p = 4209.1 - 132.8 \cdot 10^{-2}T + 143.2 \cdot 10^{-4}T^2 \tag{2.8d}$$

$$\beta = (0.8T^{0.5348} - 1.9114) \cdot 10^{-4} \tag{2.8e}$$

where T (°C) refers to the temperature of the water and ρ (kg/m^3), υ (m^2/s), k (W/mK), C_P (J/kg·K), and β (1/K) refer to its density, kinematic viscosity, thermal conductivity, specific heat capacity, and volumetric coefficient of thermal expansion, respectively.

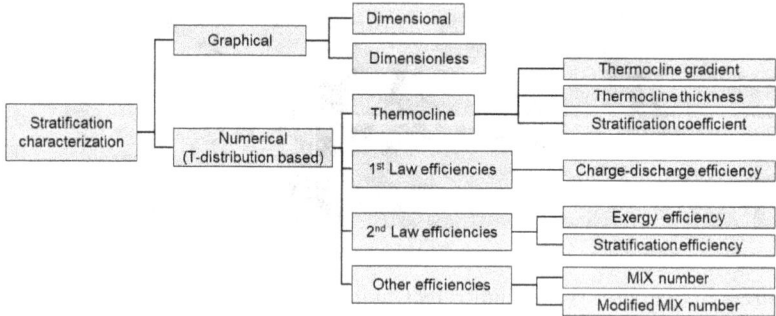

Figure 2.7. Classification of stratification characterization methods as defined by Fertahi *et al.* [14].

An appropriate hot water storage, for instance, can create large temperature differences resulting in strong thermal stratification. Additionally, kinematic viscosity decreases with increasing temperatures. Therefore, water can be transported easier in hot water stores at higher temperatures than at low temperatures. The key trade-off, however, is that thermal conductivity of water increases at higher temperatures, allowing temperature differences to equalize faster at higher temperatures [13].

Fertahi *et al.* [14] classify stratification characterization methods into two main forms as portrayed in Figure 2.7.

While dimensional graphical methods of representing thermal distribution within the tank are usually adequate for qualitative observations during the TES operation, they are unable to provide insights into the understanding and overall implications of the TES behaviour. To this end, dimensionless parameters that are used to characterize the stratification are often employed. Castell *et al.* [7] listed a few of these key dimensionless parameters that had varying significance in defining the stratification within TES units.

2.1.2.1.   *MIX number*

The MIX number was proposed and defined by Davidson *et al.* [15]. It is derived from the energy and temperature distribution levels in the LTES tank, based on the first moment of energy. Andersen *et al.* [11] further modified the number to extend the number to account for time-dependent behaviour. The modified MIX number by Andersen *et al.*

is given as

$$\text{MIX} = \frac{M_{E,\text{stratified}} - M_{E,\text{actual}}}{M_{E,\text{stratified}} - M_{\text{fully mixed}}} \tag{2.9a}$$

$$M_E = \sum_{i=1}^{n} E_i \cdot y_i \tag{2.9b}$$

$$E_i = \rho V_i C_P T_i \tag{2.9c}$$

The MIX number takes values between 0 and 1, where MIX = 0 represents a perfectly stratified tank and MIX = 1 is a fully mixed tank.

2.1.2.2. *Richardson number*

The Richardson number is used extensively in the literature to define stratification in storage tanks but more commonly in oceanic studies to predict bulk mixing occurrences [16]. The number was later found to be also useful in characterizing stratification in storage tanks by Jordan and Furbo [17]. It is employed to measure the ratio of buoyancy forces to mixing forces, broadly described [7] as

$$\text{Ri} = \frac{g\beta H(T_{\text{top}} - T_{\text{bottom}})}{v^2} \tag{2.10}$$

where g (m/s^2) refers to the gravitational acceleration, T_{top} and T_{bottom} (°C) refer to the temperature at the top and bottom of the storage tank, respectively, H refers to the height of the storage tank, and v (m/s) refers to the velocity of fluid flowing into the tank. A large Richardson number indicates a stratified tank, while a small Richardson number indicates a mixed tank. The use of this number, however, is widely varied in the literature due to the wide-ranging contexts of TES studies. Rendall *et al.* [16] summarized the diversity of interpretations, as shown in Figure 2.8.

2.1.2.3. *Peclet number*

The Peclet number describes the relationship between bulk heat transfer and conductive heat transfer, and is often used in tandem with the Richardson number to define stratification in storage

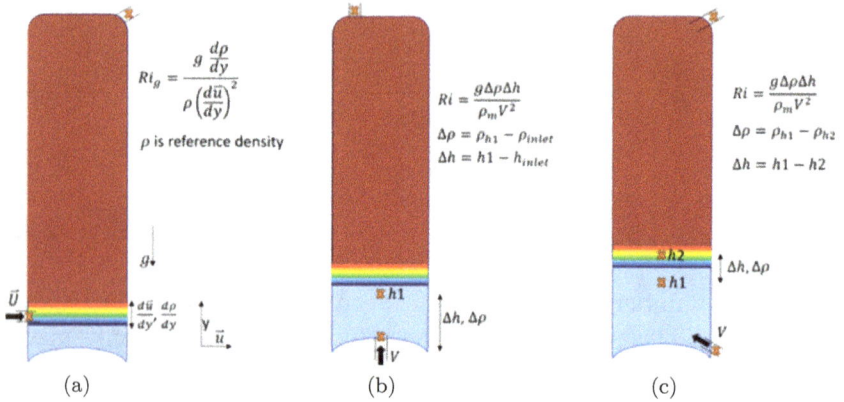

$$Ri_g = \frac{g\frac{d\rho}{dy}}{\rho\left(\frac{d\bar{u}}{dy}\right)^2}$$

ρ is reference density

$$Ri = \frac{g\Delta\rho\Delta h}{\rho_m V^2}$$
$$\Delta\rho = \rho_{h1} - \rho_{inlet}$$
$$\Delta h = h1 - h_{inlet}$$

$$Ri = \frac{g\Delta\rho\Delta h}{\rho_m V^2}$$
$$\Delta\rho = \rho_{h1} - \rho_{h2}$$
$$\Delta h = h1 - h2$$

(a)   (b)   (c)

Figure 2.8.   Common interpretations of Richardson number with (a) strictest definition to (b) superficial use for inflow from below thermocline and (c) arbitrary inlet flow direction with right being a quasi-local Ri evaluated at thermocline [16].

tanks [7]

$$\mathrm{Pe} = \frac{vH}{\alpha} \tag{2.11}$$

where α (m^2/s) refers to the thermal diffusivity of the fluid through the tank.

2.1.2.4.   *Reynolds number*

The Reynolds number indicates a flow's laminarity or turbulence, based on the ratio of inertial to viscous forces. When applied in contexts to fluid flow through pipes to packed beds, the Reynolds number of a hot or chilled water storage tank can be simply described as

$$\mathrm{Re} = \frac{\rho v D}{\mu} \tag{2.12}$$

where D refers to the entry diameter of the diffuser and μ (Pa·s) refers to the dynamic viscosity of the fluid. Rendall *et al.* [16] specified a round-jet diffuser in a storage tank to be turbulent for Re > 200, though this appears to be exceeded in most applications.

2.1.2.5.   *Stratification number*

One other less well-known dimensionless number is the stratification number, which was first introduced by Fernández-Seara *et al.* [18].

The number defines the instantaneous ratio of the mean temperature gradient to the maximum temperature gradient during the charge or discharge process, expressed as

$$\text{Str} = \frac{\left(\frac{\delta T}{\delta y}\right)_t}{\left(\frac{\delta T}{\delta y}\right)_{\text{max}}} \tag{2.13a}$$

$$\left(\frac{\delta T}{\delta y}\right)_t = \frac{1}{N-1}\left[\sum_{i=1}^{N-1}\frac{T_{i+1}-T_i}{\Delta h}\right] \tag{2.13b}$$

$$\left(\frac{\delta T}{\delta y}\right)_{\text{max}} = \frac{T_{\text{max}}-T_{\text{in}}}{(N-1)\cdot\Delta z} \tag{2.13c}$$

A summary of dimensionless numbers used for TES stratification by Rendall *et al.* [16] is detailed in Table 2.1.

2.1.2.6. *The mixing phenomena*

Two types of instabilities are induced during a charge/discharge process in most stratified STES systems, namely, Kelvin–Helmholtz (KH) and Rayleigh–Taylor (RT) instabilities. RT mixing occurs due to the gravitational effects of colder–heavier fluid interacting with a hotter–lighter fluid or vice versa. This leads to a vertical mixing process. KH mixing, on the other hand, arises from internal waves between the stratified layers. In such instabilities, incoming fluid impinges on the tank walls or internal surfaces, forming negatively buoyant plumes that interact with the thermocline. When the interaction between KH and RT instabilities that occur simultaneously cannot be easily distinguished, the process is termed "entrainment mixing" (ET). Figure 2.9 illustrates the various types of instability that occur when the fluid is introduced to a stratified TES via a horizontal inlet device.

2.1.3. *Numerical methods for stratified TES systems*

A key drawback of many of the above-mentioned methods for characterizing stratification and thermal distribution of TES tanks is that they generally rely on experimental data. This bodes poorly for

Table 2.1.   Summary of dimensionless numbers that are used to characterize TES stratification [16].

Dimensionless number	Relationship	Remarks
Biot	Fluid surface convection vs conductivity of liquid/solid	Typically used for uninsulated tanks or tanks with high-conductivity fluids
Froude	Inertial forces vs gravitational force	Inverse of the Richardson number and only smaller in scale
Grashof	Buoyancy force vs viscous force	Typically used to quantify heat loss on the walls of the tank or as an intermediate step to calculating the Richardson number for stratification analysis
Peclet	Convective heat transfer vs conductive heat transfer	Used when the fluid has a high thermal conduction or for diffusion of chemical species
Rayleigh	Thermal diffusion vs thermal convection	Criteria to determine laminar flow or critical value for mixing using a characteristic length
Reynolds	Inertial force vs viscous force	Describes the bulk flow within the tank or turbulence in the inlet device
Richardson	Buoyancy force vs momentum force	Stability of stratification in TES tanks based on buoyancy flows within the tank
Atwood	Fluid density difference vs sum of fluid densities	Used to identify the strength of the thermocline based on density difference
Stratification number	Local thermal gradient vs overall thermal gradient	Used to identify the strength of the thermocline based on temperature difference

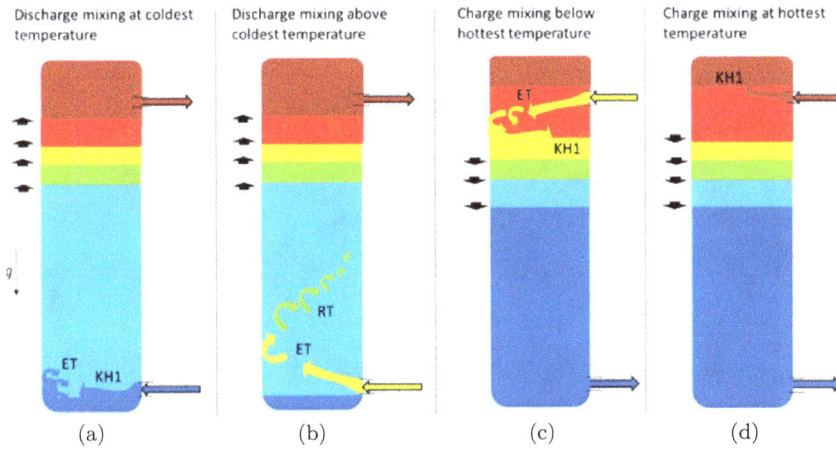

Figure 2.9. Rendall *et al.* [16] indicated the various impacts of horizontal inlet diffuser instabilities on thermal stratification such as (a) ET and KH instabilities when cold water enters at the bottom, (b) ET and RT warm mixing occurs upon entering cold bulk fluid at bottom, (c) ET and KH instabilities when warm inlet fluid at top is below T_H, and (d) when entering temperature at the top is at T_H.

designing and sizing of TES storage systems as most experimental studies only provide limited understanding of specific conditions. As a result, most techno-economic analyses and design work for TES systems today tend to rely on numerical methods instead [19].

2.1.3.1. *Single-media STES systems*

For stratified STES systems containing only a single storage medium, such as in conventional water storage tanks, basic mass, momentum, and energy conservation principles in cylindrical coordinates are often applied. To account for buoyancy effects in the working fluid, the Boussinesq approximation is also applied for axial momentum conservation. This assumes that only the density of the storage medium varies with temperature changes [20]. Commonly adopted assumptions include the following:

(1) The working fluid/storage medium is incompressible and Newtonian;
(2) Uniform initial temperature of working fluid;

(3) Isotropic and constant thermophysical properties of the working fluid over the operating temperature range; and

(4) Negligible radiative heat transfer.

$$\frac{1}{r}\left(\frac{\delta r u_r}{\delta r}\right) + \frac{\delta u_z}{\delta z} = 0 \tag{2.14}$$

$$\rho_f\left(\frac{\delta u_r}{\delta t} + u_r\frac{\delta u_r}{\delta r} + u_z\frac{\delta u_r}{\delta z}\right)$$

$$= -\frac{\delta P}{\delta r} + \mu_f\left(\frac{1}{r}\frac{\delta}{\delta r}\left(r\frac{\delta u_r}{\delta r}\right) - \frac{u}{r^2} + \frac{\delta^2 u_r}{\delta z^2}\right) \tag{2.15a}$$

$$\rho_f\left(\frac{\delta u_z}{\delta t} + u_r\frac{\delta u_z}{\delta r} + u_z\frac{\delta u_z}{\delta z}\right)$$

$$= -\frac{\delta P}{\delta z} + \mu_f\left(\frac{1}{r}\frac{\delta}{\delta r}\left(r\frac{\delta u_z}{\delta r}\right) - \frac{u}{r^2} + \frac{\delta^2 u_z}{\delta z^2}\right) + \rho_f g\beta_f(T_f - T_{fi}) \tag{2.15b}$$

$$\rho_f c_p\left(\frac{\delta T_f}{\delta t} + u_r\frac{\delta T_f}{\delta r} + u_z\frac{\delta T_f}{\delta z}\right) = \frac{1}{r}\frac{\delta}{\delta r}\left(k_f r\frac{\delta T_f}{\delta r}\right) + \frac{\delta}{\delta z}\left(k_f\frac{\delta T_f}{\delta z}\right) \tag{2.16}$$

where the subscripts r and z refer to the radial and axial planes, respectively, and the subscript f refers to the storage fluid. u (m/s), P (Pa), and T (°C) refer to the velocity, pressure, and temperature of the fluid, respectively. Figure 2.10 displays the discretization process.

Based on the assumed operating conditions, the boundary conditions imposed on the numerical model are described as

$u_w = 0$ No slip at wall condition   (2.17)

$T_w = T_{\text{insulation}}$

                         Coupled thermal boundary condition   (2.18)

$q_w = q_{\text{insulation}}$

$$-k_f\frac{\delta T}{\delta r}\bigg|_{r=0} = 0$$

                         Axisymmetric boundary condition   (2.19)

$$\frac{\delta u}{\delta r} = 0$$

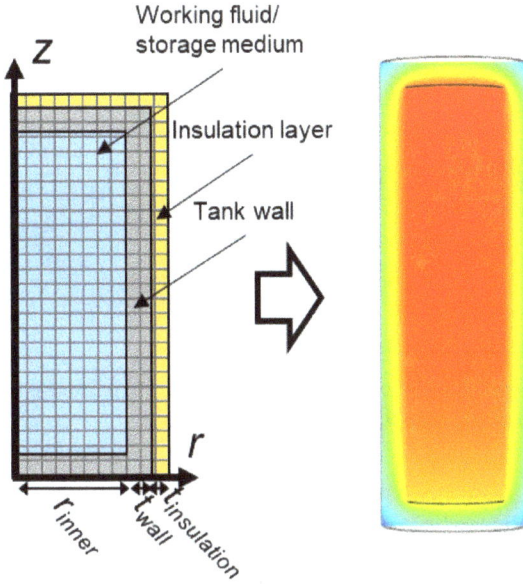

Figure 2.10. The cylindrical TES tank as an axisymmetric model and the resulting simulation result as obtained by Khurana *et al.* [20]. The model is discretized as a 2D grid.

$$-k\frac{\delta T_{\text{insulation}}}{\delta r} = h(T_{\text{insulation}} - T_0) \quad \text{Storage tank sidewall boundary} \quad (2.20)$$

$$-k\frac{\delta T_{\text{insulation}}}{\delta z} = h(T_{\text{insulation}} - T_0) \quad \text{Storage tank top and bottom boundary} \quad (2.21)$$

where T_0 (°C) refers to the ambient temperature and the subscripts w and "insulation" refer to the TES wall and insulation, respectively. Lastly, heat loss to the ambient is computed via the Nusselt number formulation as

$$\overline{Nu}^{\frac{1}{2}} = 0.825 + \frac{0.387 Ra^{\frac{1}{6}}}{\left[1 + \left(\frac{0.437}{Pr}\right)^{\frac{9}{16}}\right]^{\frac{8}{27}}} \quad (2.22)$$

where Ra and Pr refer to the Rayleigh and Prandtl numbers of the surrounding air, respectively.

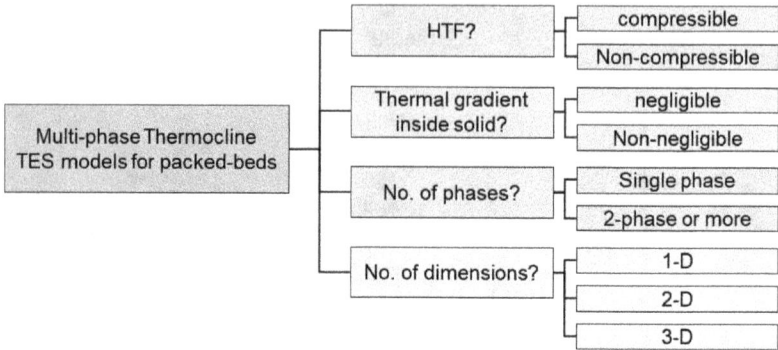

Figure 2.11.   Classification of numerical methods for packed-bed STES systems by Palombo and Frazzica [19].

2.1.3.2.   *Multi-media STES systems*

Single-media STES systems such as chilled/hot water storage tanks constitute the majority of urban and domestic use cases. Recently, multi-media systems that involve more than one storage material have become more prevalent, especially in high-temperature applications such as concentrating solar power (CSP). These often take the form of packed-bed TES systems that consist of a liquid or gaseous heat transfer fluid (HTF) with a solid such as a quartzite rock bed. Palombo and Frazzica [19] classify commonly used multi-phase numerical models for such packed beds as displayed in Figure 2.11. Generally, simplified models that represent the entire TES as a single phase require less overall computational power at the expense of some degree of accuracy. As more layers of complexity, such as internal thermal gradients and HTF compressibility, are introduced, the required computational intensity is considerably increased. The number of dimensions designated in the model also has a marked impact on the computational power required. It is noteworthy that some well-simplified 1D or 2D models with the appropriate assumptions can potentially present simulated results as accurately as 3D models. These are summarized in Table 2.2.

1D models are the simplest to develop and solve, and start from the assumption that radial thermal variation is negligible relative to axial thermal distribution within the TES. 1D models often assume adiabatic wall regions with separate solid and liquid phases, in which

Table 2.2.   Impact of different models for packed-bed thermocline TES on computational power required [19,21].

Model type	CPU time (s)
1D single phase	7.1
1D single phase (continuous solid)	51.6
2D single phase	141.5
2D single phase (continuous solid)	127.7
2D/3D multi-phase with thermal gradient	574.6

the heat equations are solved separately. A porous system of void ratio equivalent to the porosity of the packed bed is also assumed to compute the heat transfer between the HTF and solid. The 1D energy equation is then written as follows:

$$\frac{\delta}{\delta t}(\varepsilon\rho_f c_{pf}T_f) + c_{pf}\frac{\delta}{\delta z}(\rho_f u_f T_f) - \frac{\varepsilon\delta p}{\delta t}$$
$$= \frac{\delta}{\delta z}\left(k_{\text{eff}}\frac{\delta T_f}{\delta z}\right) - h\alpha(T_s - T_f) + h_w S_w(T_w - T_f) \qquad (2.23)$$

where subscripts s and f refer to the TES solid material and wall, respectively, and ε refers to the packed bed porosity factor. The left-hand side of the equation accounts for the axial thermal gradient and pressure differential across the packed bed. The right-hand side of the equation includes the source term $h\alpha(T_s - T_f)$ which accounts for the heat transfer between the HTF and the solid material, where h refers to the corresponding heat transfer coefficient. The solid material energy equation is then formulated as

$$(1-\varepsilon)\frac{\delta}{\delta t}(\rho_s c_{ps}T_s) = \frac{\delta}{\delta z}\left(k_{\text{eff}}\frac{\delta T_s}{\delta z}\right) - h\alpha(T_s - T_f) \qquad (2.24)$$

Additionally, the continuity equation through the packed bed is formulated as

$$\varepsilon\frac{\delta\rho_f}{\delta t} + \frac{\delta}{\delta z}(\rho_f u) = 0. \qquad (2.25)$$

The heat transfer coefficient, h, is often deduced through commonly used correlations as detailed in Table 2.3. The pressure drop is

Table 2.3.   Commonly used Nusselt number correlations for packed beds [19].

Correlation	Authors/Reference	Remarks
$\mathrm{Nu} = 3.22\,\mathrm{Re}^{\frac{1}{3}}\mathrm{Pr}^{\frac{1}{3}}$ $+0.117\,\mathrm{Re}^{0.8}\mathrm{Pr}^{0.4}$	Beek [23]	Randomly arranged particles
$\mathrm{Nu} = 2 + 1.1[6(1-\varepsilon)]^{0.6}$ $\times \mathrm{Re}^{0.6}\mathrm{Pr}^{\frac{1}{3}}$	Wakao and Funazkri [24]	Most commonly used, porosity-dependent
$\mathrm{Nu} = 2 + 2.03\,\mathrm{Re}^{0.5}\mathrm{Pr}^{\frac{1}{3}}$ $+0.049\mathrm{RePr}^{0.5}$	Galloway and Sage [25]	
$\mathrm{Nu} = \mathrm{Re}^{0.5}\,\mathrm{Pr}^{0.33}$	Bedecarrats *et al.* [26]	Suitable for dominant forced convection
$\mathrm{Nu} = 18.1\,\mathrm{Pr}^{\frac{1}{3}}$	Vafai and Sozen [27]	Suitable for Re < 50
$\mathrm{Nu} = 2 + \dfrac{0.589\mathrm{Ra}^{0.25}}{\left[1+\left(\frac{0.469}{\mathrm{Pr}}\right)^{\frac{9}{16}}\right]^{\frac{4}{9}}}$	Churchill	Suitable for dominant natural convection

frequently quantified via the Ergun equation described as

$$\Delta P = A\frac{(1-\varepsilon)^2 \mu}{\varepsilon^2 d^2}u_f + \frac{B(1-\varepsilon)\rho_f}{\varepsilon^3 d}u_f^2 \qquad (2.26)$$

where d (m) refers to the diameter of the solid particles in the packed bed, and A and B are empirical constants. For spherical particles, A and B are 150 and 1.75, respectively [22].

2D models can be broadly classified into single-phase or 2-phase models. In single-phase 2D models, a single temperature gradient is considered for both solid and fluid domains. For 2-phase models, they consider two separate thermal gradients for each domain. The mass and momentum conservation equations are formulated as

$$\frac{\delta(\varepsilon\rho_f)}{\delta t} + \nabla \cdot (\rho_f \vec{u}) = 0 \qquad (2.27)$$

$$\rho_f\left(\frac{\delta\vec{u}}{\delta t} + \nabla \cdot \frac{\vec{u}(\vec{u})}{\varepsilon}\right) = -\varepsilon\nabla p + \varepsilon\rho_f \vec{g}$$

$$-\varepsilon\left(\frac{\mu_f}{K}\vec{u} + \frac{\rho_f F}{\sqrt{K}}|\vec{u}|\vec{u}\right) \qquad (2.28)$$

$$K = \frac{\varepsilon^3 d^2}{36\kappa(1-\varepsilon)^2} \qquad (2.29)$$

where K and F refer to the permeability and inertial coefficient of the bed, respectively. The permeability of the bed can be derived from the Kozeny–Carman equation as detailed in Equation (2.29) and κ is the Kozeny–Carman constant, which is taken to be 5 for spherical particles. The final term on the right-hand side of Equation (2.28) denotes a "momentum source" that is added to model the fluid flow in porous media. The $\varepsilon \rho_f \vec{g}$ term describes viscous dissipation, where g refers to the gravitational acceleration. In single-phase models, the key assumption is that the solid and liquid phases are under local thermal equilibrium, which is mostly only valid under specific circumstances such as having a Biot body where the heat transfer is high between the solid and liquid phases. This is prevalent in cases where the particle size of the solid filler in the TES is small compared to the TES size or when the HTF flow is laminar. Accordingly, the energy equation can be formulated as

$$\frac{\delta}{\delta t}(\varepsilon \rho_f c_p T + (1-\varepsilon)\rho_s c_{ps} T) + \nabla \cdot (u(\rho_f c_{pf} T + p\rho))$$
$$= \nabla \cdot (k_{\text{eff}} \nabla T) + h_w S_w (T_w - T_f) \qquad (2.30)$$

where k_{eff} (W/mK) refers to the effective thermal conductivity of the bed.

In 2-phase models, two separate equations for the solid and liquid domains are formulated instead, as follows:

$$\frac{\delta}{\delta t}(\varepsilon \rho_f c_p T_f) + \nabla \cdot (\vec{u}(\rho_f c_{pf} T_f))$$
$$= \nabla \cdot (k_{\text{eff}} \nabla T_f) + h\alpha(T_s - T_f) + h_w S_w (T_w - T_f) \qquad (2.31)$$

$$\frac{\delta}{\delta t}((1-\varepsilon)\rho_s c_{ps} T_s)) = \nabla \cdot (k_{\text{eff}} \nabla T_s) - hS(T_s - T_f) \qquad (2.32)$$

2.1.3.3. *Solving the models computationally*

The aforementioned models are solved either through commercial software or in-house developed programmes. The most commonly used software include ANSYS Fluent® and COMSOL, especially for 2D models where there are significant levels of complexity involved. In contrast, 1D models are commonly solved using MATLAB. In-house codes developed include the Three Diagonal Matrix Algorithm (TDMA) and the Alternating Direction Implicit (ADI) scheme.

The NesT code is sometimes used in tandem with the TDMA to link multiple different components. A modified SIMPLE algorithm, named the SIMPLER algorithm (Semi-Implicit Method for Pressure–Linked Equations Revised), has also been developed for pressure–velocity coupling.

2.2.   STES Design and Optimization

Most often, the primary objective of a TES study is to arrive at the most optimal design and sizing of a TES unit for a specific application. However, this endeavour is often non-trivial. A typical approach taken entails carrying out a parametric study using a computational model based on data from experimental testing. Once the sensitivity of the selected TES design parameters on the key TES performance metrics is determined, the optimal design can then be derived. Presently, this is often performed by employing meta-heuristic optimization schemes implemented through appropriate computational codes. The fundamental deficiency of this approach, despite implementing in the wide range of TES studies, is that the diversity of TES configurations limits the versatility of the generalized formulations for TES design optimization. As a result, the study findings become highly specific with little generalized knowledge, thereby, reducing the contributing values to the field.

2.2.1.   *Modelling and methods*

While the methods described in Section 2.1 are generally applicable for most computations and performance prediction studies, many other methods have also been tested and evaluated in the literature. Baeten *et al.* [28] presented a basic 1D finite volume discretization method for a simple water storage tank coupled with interlayer mixing and buoyancy formulations. This was executed by representing an energy flowrate caused to natural convection and inflow mixing as a mixing flow over each liquid layer in the tank as depicted in Figure 2.12. The distribution is determined by two parameters: (1) the temperature difference between the mixing layer and the average temperature across all layers, and (2) the position of the mixing layer relative to the inflow. Cross-layer mixing that is caused

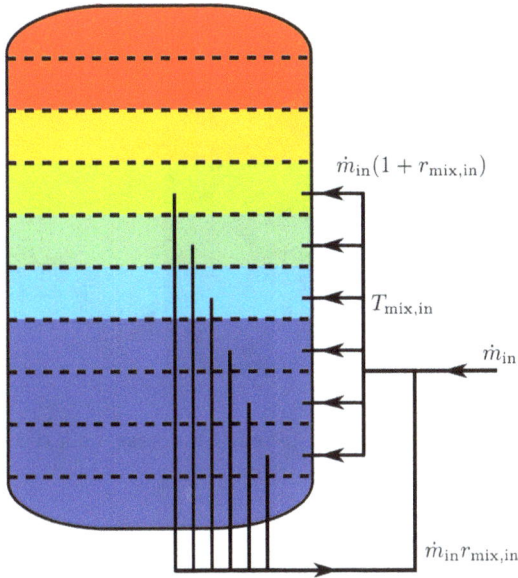

Figure 2.12. Buoyancy and inflow mixing model coupled with 1D finite volume scheme for storage tank by Baeten *et al.* [28].

by entrainment of fluid in the incoming buoyant jet is also accounted for. Temperature of the mixing flow is deduced by the mixing fluid from each layer and the inflow according to the distribution amount. The model was computed using the Modelica language on *Dymola 16* software. A separate CFD model is also designed using CAD and tested in ANSYS with both the k–ε and k–ε–SST turbulence models.

Lastly, a non-dimensional storage efficiency index termed the effective capacity ratio is defined as the ratio of the full-discharge time period to the full-charge period for the same charge and discharge power. This ratio allows the induced mixing inefficiencies to be normalized according to the flow rate, which is a key determinant of the thermocline. Consequently, an effective capacity ratio of 1 denotes a perfectly stratified tank where no mixing has occurred as the charge and discharge periods are considered to be equivalent. This index, therefore, expresses the actual storage capacity against that of a perfectly stratified tank.

Model validation is often carried out using the experimental data acquired from a physical setup as depicted in Figure 2.13. 34 T-type thermocouples were installed in the $0.3\,\mathrm{m}^3$ storage tank, divided into

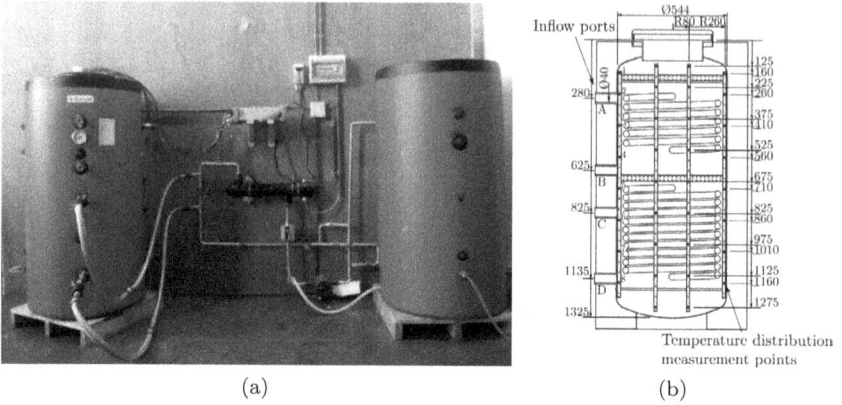

(a)   (b)

Figure 2.13.   (a) Experimental setup and (b) storage tank drawing and instrumentation points [28].

4 vertical arrays. The storage tank is supplied with hot water from a second identical tank, as shown in Figure 2.13.

Based on CFD simulations, a linear relation between the mixing ratio (defined as the ratio of the mixing flow rate per layer to the incoming flow rate) and Reynolds number can be established at both inflow and outflow. The behaviour of the inflow and outflow characteristics for a strong-jet and plume-like inflow, simulated by CFD, is as shown in Figure 2.14.

Comparisons made between the proposed model and the CFD model validated against experimental data has revealed that the proposed model is able to reduce computation time by more than 500,000 times while providing comparably equivalent results. Key sources of deviation from the data provided by the model are attributed to the uniform flow distribution assumption. Furthermore, the findings confirmed the existence of an approximate linear correlation between the inflow mixing zone and the effective storage capacity ratio as illustrated in Figure 2.15. Apparently, less inflow mixing results in a better effective storage capacity ratio. This finding highlights the importance of inflow mixing modelling as well as the proposed model in performance prediction of the storage.

Rahman and Smith [29] took a different sizing approach by incorporating Neural Networks to optimize the design of storage tanks. Machine Learning (ML) methods are often utilized by researchers and professionals alike in predicting future loads based on past data.

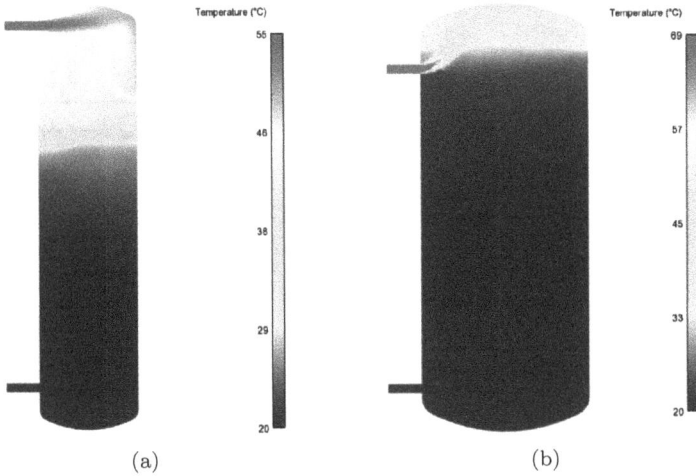

Figure 2.14. TES behaviour for the (a) strong-jet case and (b) plume-like case as reflected from the CFD study [28].

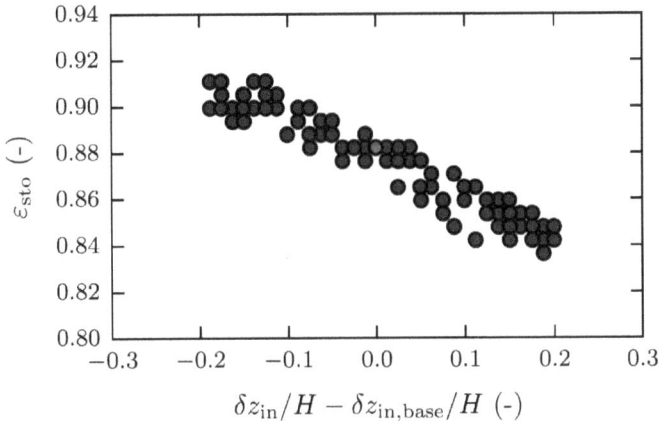

Figure 2.15. Correlation between the inflow mixing zone thickness (horizontal axis) and the effective storage capacity ratio (vertical axis) [28].

Simple algorithms from basic multivariate linear regression to multi-layered perceptron (MLP) neural networks (NN) and Gaussian processes have been widely employed to predict building thermal and electrical loads. A deep Recurrent Neural Network (RNN) model has been used to predict the hourly heating load in multiple campus buildings in the University of Utah in Salt Lake City. Input

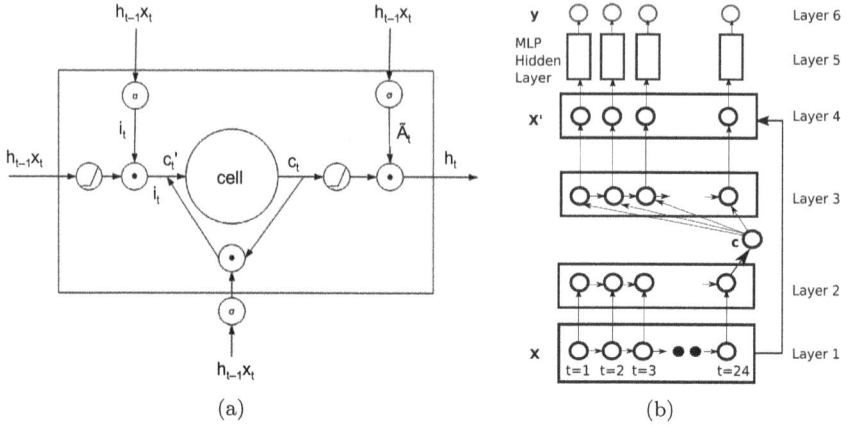

Figure 2.16.   (a) Schematic representation of long short-term memory (LSTM) activation function and (b) schematic representation of the proposed RNN model. Layers 2 and 3 of the RNN model contain LSTM to model temporal dependencies and are analogous to encoder–decoder architecture [29].

variables for the RNN model included weather such as psychrometric data, as well as the calendar date and time down to the hour. The proposed RNN model uses an encoder–decoder architecture that leverages on the periodicity of electrical consumption, as shown in Figure 2.16(b).

The layers of the RNN model are briefly described as follows, based on [29]:

1. Layer 1 contains the input features X for 1 h every 24 h
2. Layer 2 converts the inputs into a fixed vector representation

$$X \in R^{24 \times d}$$

3. Layer 3 receives the fixed vector from the second layer and generates vector at each timestep, which acts as a surrogate for scheduling/time variables

$$c = h_{t,e} = \text{LSTM}(h_{t-1,e}X)$$

4. Layer 4 concatenates the surrogates and inputs and ensures dependencies remain

$$h_{t,d} = \text{LSTM}(ch_{t-1,d})$$

5. Layer 5 introduces the concatenated input X' to a shared MLP layer

$$X' = [X; h_{t,d}]$$

6. Layer 6 contains the hourly heating demand in 24-h sequences

$$y_{p,t=1,2,3,...,24\,\text{h}} = \text{MLP}(X'_{t=1,2,...,4\,\text{h}})$$

The model is then regulated with weight decay regularization and early stopping to prevent over-fitting. The model has been developed using the *Keras* API in Python on a *Theano* backend. The accuracy of the prediction outcomes has been evaluated using root-mean-squared error (RMSE). The storage tank itself, however, is modelled as a simple 1D model comprising a cylindrical tank with two embedded heat exchangers as portrayed in Figure 2.17(a). The model assumed the tank to be adiabatic and the system to be closed. Incompressible flow and a high systemic pressure greater than 0.5 MPa is assumed. Additionally, radial temperature variation is neglected.

The heat transfer coefficient between the heat exchangers and the storage tank has been deduced through a thermal circuit formulated as

$$\text{UA} = \frac{1}{\left[\left(\dfrac{1}{h_i A_i} + \dfrac{\ln\left(\frac{d_o}{d_i}\right)}{2\pi k \Delta x} + \dfrac{1}{h_o A_o}\right)\right]}$$

where h, A, and d refer to the convective heat transfers, surface areas, and diameters of the heat exchanger, and the subscripts i and o refer to the inner and outer surfaces, respectively. The overall heat balance equations are solved via an implicit scheme in MATLAB using a time-step of 10 s for 10 tank layers (or "nodes"). The optimization process involves a Bayesian optimization scheme based on the *bayesopt* package developed in Python.

Predictions made by the RNN model and a reference MLP model have been compared against actual heating data over 120 h for each test site on campus. Figure 2.18 shows that the RNN model proposed is found to be much more consistent than the MLP model in many cases. The RMSE value relative to the root mean squared average of the actual building heating demand was found to be as low as 0.271

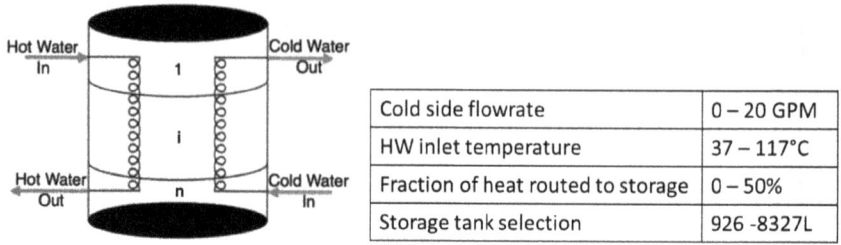

(a)

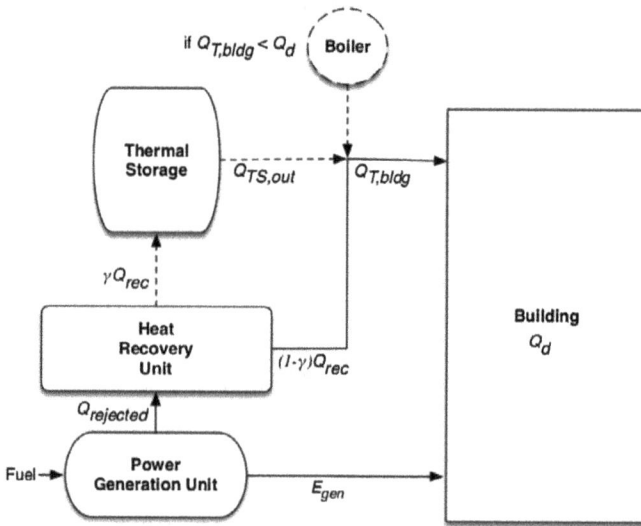

(b)

Figure 2.17.   (a) Model of TES tank with embedded cold and hot side exchangers as well as multi-layer modelling with the search space for the associated parameters, and (b) schematic representation of thermal distribution network in the campus [29].

for the RNN model but up to 0.552 for the MLP model. The proposed optimization scheme was then compared against test cases where the optimization scheme was omitted via the level of demand the TES supports against the total heating demand. Over a time scale of 59 days, the optimized scheme shows a 0.26% discrepancy in contrast to the non-optimized scenario as shown in Figure 2.19. Physically, the results mean that the proposed framework using Gaussian processes is able to utilize the deep RNN predictions to estimate the CHP-TS behaviour with improved efficiency and comparable accuracy. Key

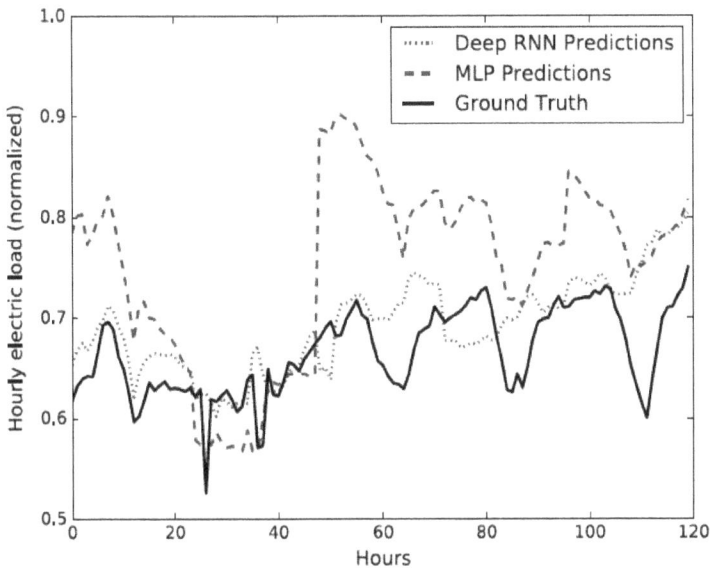

Figure 2.18. Heating demand predictions between RNN and MLP model between January 21st and 26th, 2017 compared against actual live data.

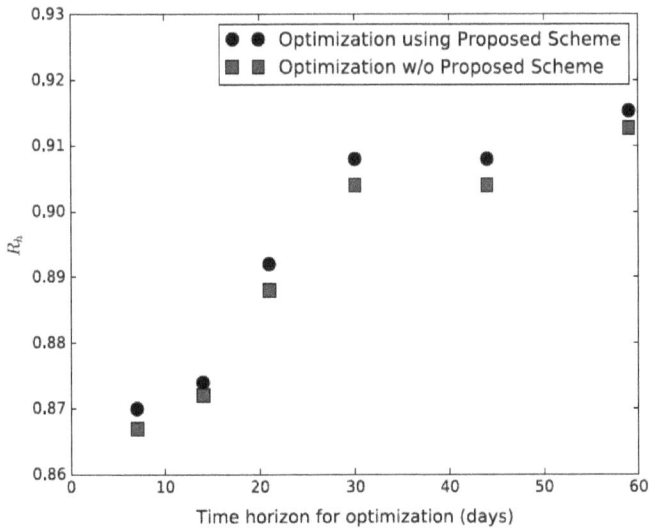

Figure 2.19. Level of TES support for heating demand compared between optimized scenarios with and without the proposed scheme [29].

findings further reveal that the proposed optimization scheme is feasible for long-term predictions for sizing of a TES for building-scale applications.

In a more recent study, Khurana *et al.* [20] employ a novel approach involving a Box-Benken Design (BBD) method followed by applying an analysis of variance method (ANOVA) to the results in order to develop a reduced formulation for predicting the internal conditions of a storage tank under standalone conditions. The numerical model of the storage tank is first constructed as per details in Section 2.1.3.1 and simulated using *COMSOL Multiphysics 5.4a* software. The transient performance of the model is subsequently validated with acquired experimental data based on the setup shown in Figure 2.20(a). Simulated results based on the defined numerical model are found to be in good agreement with experimental data, with only a maximum deviation of 2.09%, as highlighted in Figure 2.20(b) for the 50°C and 70°C initial temperature test cases. The discrepancy between the simulated and experimental data is attributed to the fluctuating ambient conditions during experiments.

In an attempt to assess the impacts of key parameters on the response of the tank, a response surface methodology (RSM) is utilized. While prior endeavours in the literature have employed one-factor-at-a-time (OFAT) techniques, this method does not account for the interactive effects between parameters, limiting its ability to predict quadratic effects without large datasets. The BBD method is then chosen as it is known to be adequately accurate with minimum iterations compared to other RSM types. The selected parameters include storage volume, tank aspect ratio, charging temperature, ambient temperature, insulation thickness, storage tank height, axial height of temperature point, and storage time ratio. The parameters are set at 3 levels (high, medium, low) as detailed in Table 2.4, and are tested for 62 runs. The predicted response is approximated by a second-order polynomial formulated as

$$Y = \beta_o + \sum_{i=1}^{k} \beta_i x_i + \sum_{i=1}^{k} \beta_{ii}(x_i)^2 + \sum_{i=1}^{k-1} \sum_{j=1}^{k} \beta_{ij} x_i x_j$$

where β_o, $\beta_i \beta_{ii}$, and β_{ij} are the regression coefficients, respectively, and x_i and x_j are the process parameters. ANOVA is then utilized to evaluate the significance of the main effects and interactions.

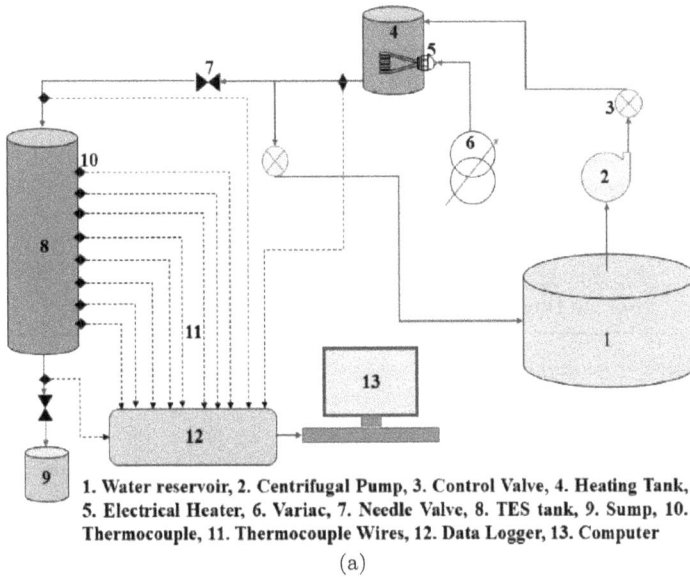

1. Water reservoir, 2. Centrifugal Pump, 3. Control Valve, 4. Heating Tank,
5. Electrical Heater, 6. Variac, 7. Needle Valve, 8. TES tank, 9. Sump, 10.
Thermocouple, 11. Thermocouple Wires, 12. Data Logger, 13. Computer

(a)

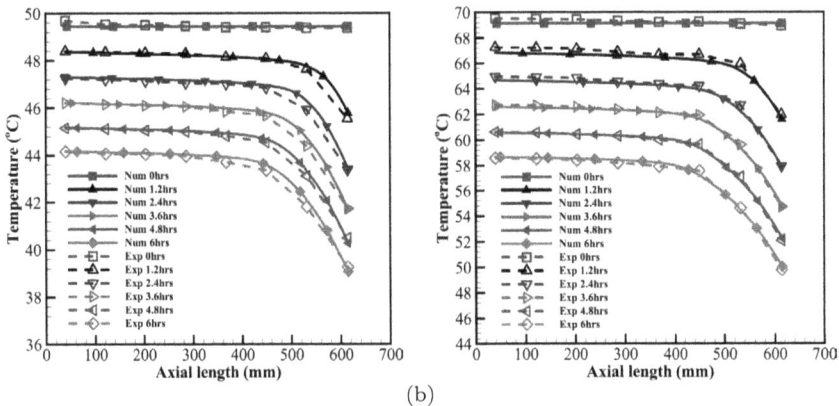

(b)

Figure 2.20. (a) Schematic representation of the experimental setup and
(b) validation of the numerical model with experimental data for the 50°C and
70°C initial temperature test cases [20].

ANOVA on the BBD results has revealed that the terms A, B, C,
D, E, F, G, AA, FF, AC, AF, AG, BF, CE, CF, CG, DF, DG, and
EG (refer Table 2.4) are found to be statistically significant based
on their respective P-values, while the other interaction terms are
found to be insignificant. The full prediction equation is expressed

Table 2.4.   Values for each "level" of parameter [20].

Label	Parameter	Units	Low	Medium	High
A	Storage volume	mm^3	0.2	0.6	1
B	Aspect ratio		2	3	4
C	Charging temperature	°C	50	70	90
D	Ambient temperature	°C	23	29.5	36
E	Insulation thickness	mm	40	60	80
F	Axial length ratio		0	0.5	1
G	Storage time ratio		0.027	0.5135	1

as follows:

$$
\begin{aligned}
T = {}& 2.14 - 0.17A + 0.133B + 0.9833C - 0.094D - 0.0135E \\
& + 3.99F - 2.15G - 2.396A^2 + 0.0779B^2 - 0.000018C^2 \\
& + 0.00238D^2 - 0.000066E^2 - 4.884F^2 + 0.105G^2 + 0.132AB \\
& + 0.036AC - 0.0011AD - 0.0017AE + 0.99AF + 2.685AG \\
& - 0.00272BC - 0.0112BD + 0.00086BE - 0.606BF - 0.136BG \\
& - 0.000076CD + 0.000396CE - 0.0677CF - 0.0852CG \\
& - 0.000463DE + 0.0821DF + 0.0823DG + 0.0136EF \\
& + 0.0183EG + 0.133FG
\end{aligned}
$$

while the reduced form after removal of the insignificant terms could be expressed as

$$
\begin{aligned}
T = {}& 2.14 - 0.17A + 0.133B + 0.9833C - 0.094D - 0.0135E + 3.99F \\
& - 2.15G - 2.396A^2 - 4.884F^2 + 0.036AC + 0.99AF + 2.685AG \\
& - 0.606BF + 0.000396CE - 0.0677CF - 0.0852CG + 0.0821DF \\
& + 0.0823DG + 0.0183EG
\end{aligned}
$$

An F-value of 2,873 indicates a high degree of model predictability with a coefficient of determination of 99.97%. A comparison between the predicted response and the actual response shows a strong agreement between the model with the numerically simulated data, as depicted in Figure 2.21.

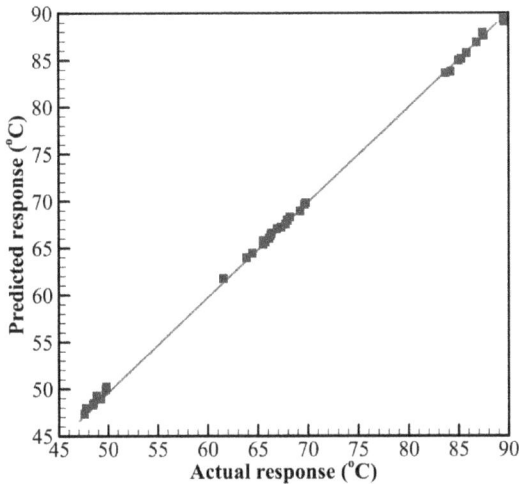

Figure 2.21. Comparison between the BBD design model prediction and numerical simulation data [20].

2.2.2. *STES tank sizing and design*

Albeit the conventional cylindrical storage tank design being the mainstay of many stratified TES systems to this day, considerable effort has been made to assess alternative forms of TES design. Yang *et al.* [30] computationally modelled and tested 10 different storage tank shapes of equal volume, material, and wall thickness. A finite-volume scheme with implicit first-order was applied to the numerical solution of the governing equations. Following the validation of the model via experimental data, each of the 10 designs was modelled similarly and tested. The design models are as shown in Figure 2.22. The initial temperature conditions of 80°C and ambient condition of 15°C were imposed on all cases, with the tank models allowed to cool to ambient temperature within 13 h. All cases were modelled as SS 304 with a storage volume of 33.6 L. Wall thickness was kept constant across all cases at 1 mm with insulation of 10 mm thick glass wool.

Thermal energy storage efficiency of the sphere and barrel were found to be the highest of the shapes at 72.69% and 72.46%, respectively, as shown in Figure 2.23(a). Contrary to common practice, the cylinder was actually found to be the most inefficient design, recording only 69.56% storage efficiency at 12 h. Yang *et al.* [30] deduced

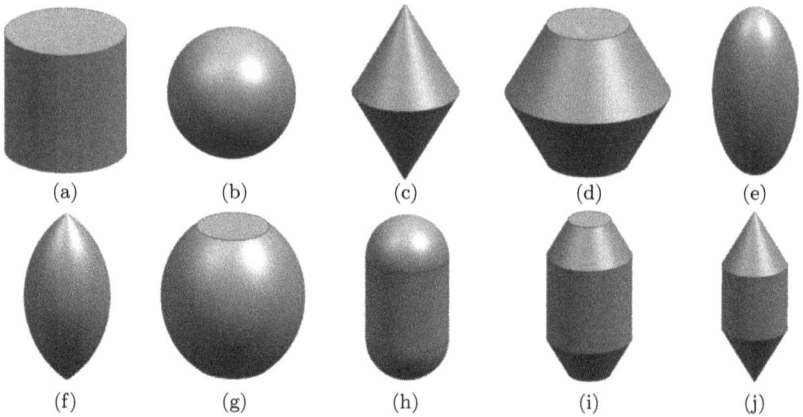

Figure 2.22.   Varying storage tank shapes as modelled and compared by Yang *et al.* [30] (a) Cylinder; (b) sphere; (c) cone; (d) truncated cone; (e) ellipsoid; (f) spindle; (g) barrel; (h) cylinder + sphere; (i) cylinder + truncated cone; and (j) cylinder + cone.

that the surface-area-to-volume ratio was the key factor affecting the thermal energy storage efficiency. This pattern was also observed for the thermal exergy storage efficiency shown in Figure 2.23(b). A simple ranking highlights that the sphere and barrel designs top the list with an energy efficiency of 54.25% and 53.93%, respectively. Contrastingly, the cylinder achieves the lowest exergy efficiency at 49.85%. It is noteworthy that the exergy storage efficiency assesses the degree of mixing of the hot and cold fluids in the tank as well as the heat loss to the environment.

A key distinction is made between the tank shapes. Based on the internal condition transformation over time, they can be classified into shapes with angled corners (consisting of spindle, cone, and cone + cylinder), hemispherical shapes (sphere, cylinder + sphere, and ellipsoid), and shapes with horizontal plane surfaces (truncated cone, barrel, cylinder + truncated cone, and cylinder). For angled-corner shapes, no flow in the stratification region is observed while flow circulation is found to occur in the isothermal region with no vortices generated. For hemispherical shapes, the flow in the stratification region is observed to be weak with a simple vortex system at the top region. In the last category, a complex vortex is observed at the top of the tank. In sum, the angled cornered shapes produced the highest degree of stratification while shapes with horizontal plane

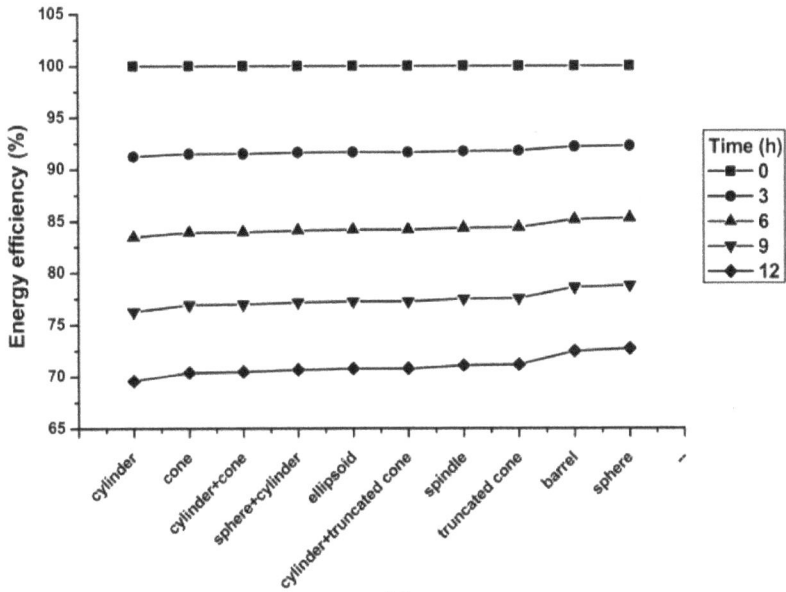

(a)

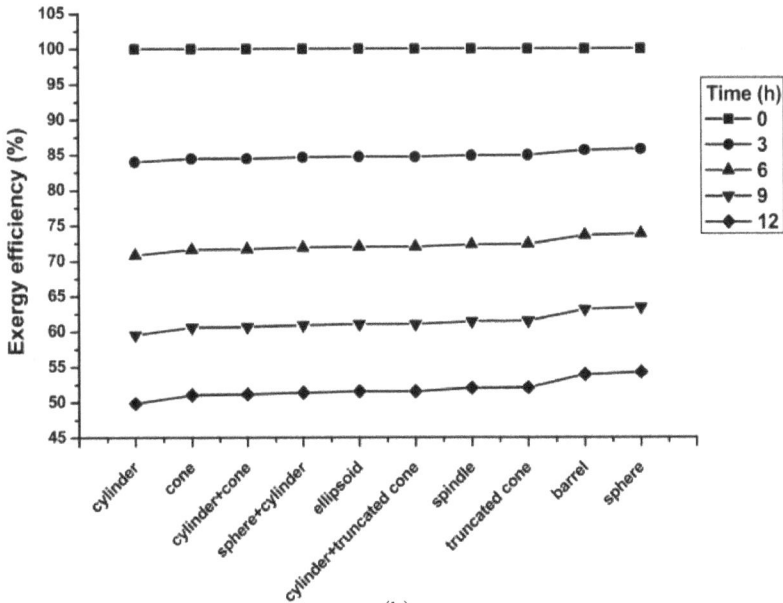

(b)

Figure 2.23. Comparison between different shaped tanks for thermal (a) energy and (b) exergy storage efficiency [30].

surfaces had the least. This difference is attributed to the stationery fluid at the bottom of the angled corner-shaped tanks, resulting in a sustained drop in temperature to the environment in this region. For tanks with horizontal plane surfaces, flow is more prominent at the bottom of the tank and thermal stratification is further induced by downward flow in the boundary layer of the lateral walls.

Sun *et al.* [31] conducted a novel study on a physically separated chilled water tank which utilizes a fabric layer insulated with 10 mm rubber as a separator to circumvent mixing problems in stratified TES systems. As shown in Figure 2.24(a), the tank had a storage capacity of up to $1.6\,m^3$ with a height and diameter of 1.2 m and 1.6 m, respectively. The tank was experimentally tested in a system consisting of a chiller and heater unit to meet and simulate the cooling load as depicted in Figure 2.24(b).

The half-FOM of the system was measured under varying operating temperatures and flow rate conditions. $FOM_{1/2}$ was observed to be higher for cases where the charging temperature was lower (2°C) compared to those with a conventional temperature of 4°C. This is highlighted in Figure 2.25. Higher temperature differences between the charge and discharge conditions also lead to higher $FOM_{1/2}$, even though the higher temperature difference does lead to higher heat transfer between the hot and cold sides. Overall, the tank appears to be able to achieve at least 92% for half-cycle and between 85.7% and 93.2% FOM for full-cycle.

2.2.3.   *Stratifiers, diffusers, and inlet devices*

One more focused and specific aspect of STES study, especially in hot or chilled water storage tanks, involves diffusers and inlet devices. The design of the inlet devices, along with stratifiers and diffusers, constitutes an important aspect in managing TES stratification. Throughout the years, different variants of TES stratifiers and diffusers have been studied and applied as shown earlier in Figures 2.5 and 2.9. Kocijel *et al.* [32] conducted a numerical study on three different radial plate diffuser designs, shown in Figures 2.26(b)–2.26(d), for hot water storage in a district heating system. The numerical model has been validated using performance data from a radial valve in a hermetic compressor, which is geometrically similar. The three designs revolve around the same principle of a radial-plate diffuser

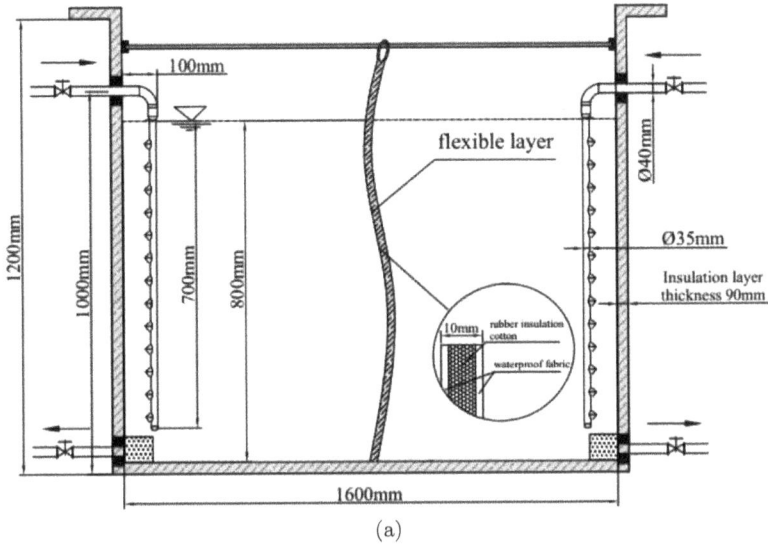

(a)

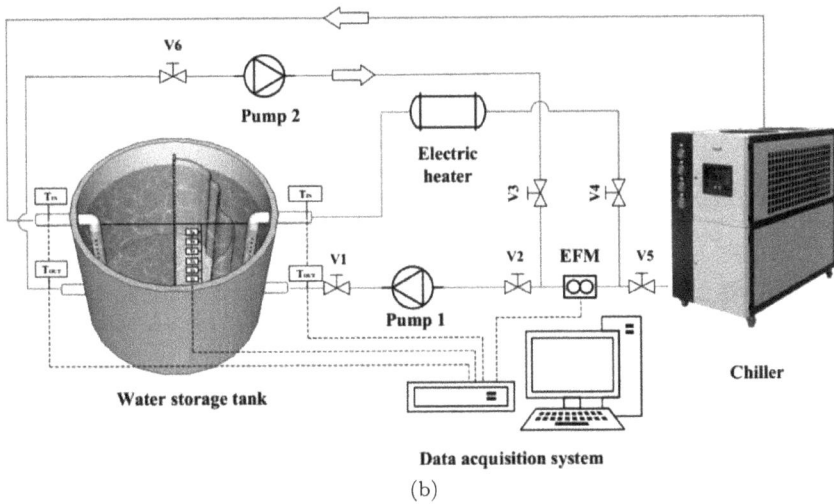

(b)

Figure 2.24. (a) Physically separated storage tank design and (b) experimental setup schematic representation [31].

where the incoming water enters the water storage body through the gap between parallel plates. Variations in the designs involve changing the joint angle connecting the incoming pipeline and the parallel plates, leading to right-angled joint (sharp edge), rounded joint (curved element), and conical joint scenarios.

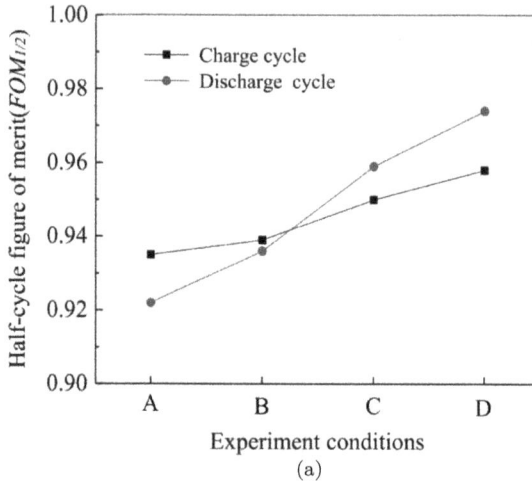

Figure 2.25.   Trend of $FOM_{1/2}$ against the various experimental test conditions conducted (a) and the operating details for each test set (A, B, C, and D) (b).

	Test	Flow(L/min)	T_{in} (°C)	T_{out} (°C)	$\triangle T$ (°C)
Charging	A-1	9.30	4.41	12.32	7.91
	B-1	8.72	2.08	12.34	10.26
	C-1	7.87	1.93	14.26	12.33
	D-1	6.78	2.08	15.94	13.86
Discharging	A-2	9.67	12.14	4.55	7.59
	B-2	7.33	12.21	2.65	9.56
	C-2	6.33	13.98	2.88	11.10
	D-2	5.50	16.18	2.96	13.22

(b)

Key findings have revealed that the sharp-edged joint (as depicted in Figure 2.26(c)) produced the highest degree of thermal stratification in the tank, while also inducing the highest static pressure drop. This is then followed by the curved-element (rounded-edge joint) design which produces slightly less stratification and static pressure drop. For this design, increasing the curvature radius leads to reducing the thermal stratification. Deng *et al.* [33] conducted a similar study on radial plate diffusers and also confirmed that the plate diameters did not have any significant impact on the TES stratification. However, some level of cost savings of up to 37.5% was still possible for unequal plate diameters (Figure 2.27).

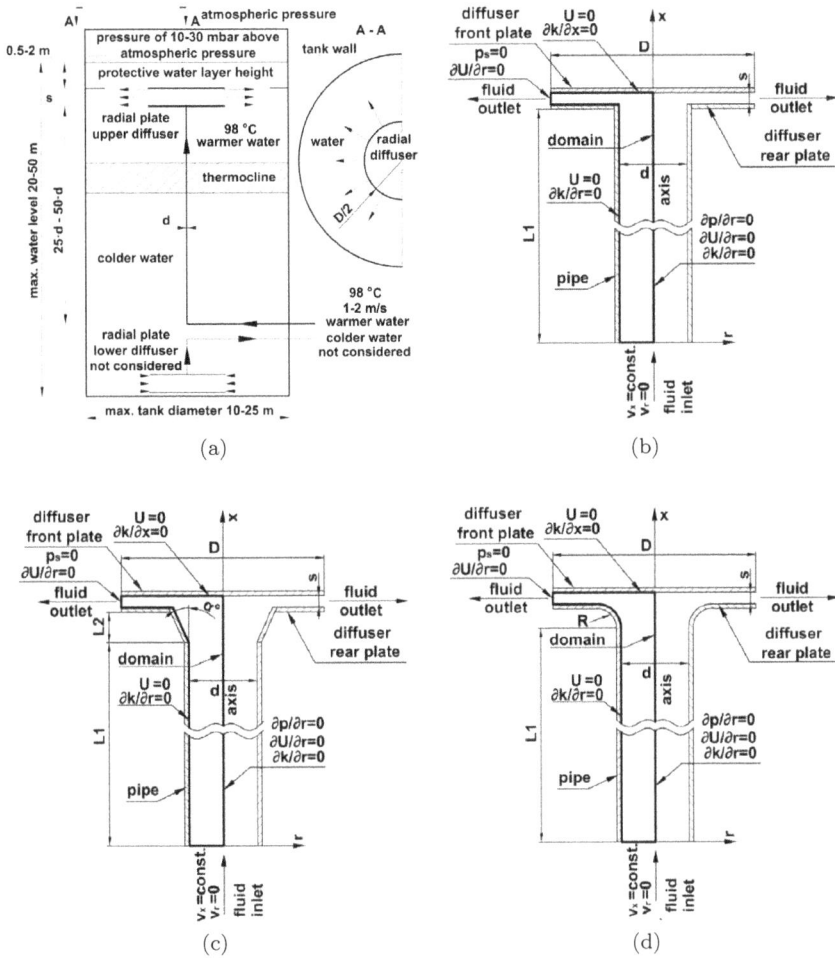

Figure 2.26. (a) The hot water TES with radial plate diffuser and the various designs; (b) sharp edge joint; (c) conical element; and (d) curved element [32].

Parida *et al.* [34] proposed a novel hemispherical diffuser design for a molten salt STES and compared its performance to 3 other cases, namely, a diffuser-less inlet, a flat-plate solid diffuser, and a coaxial ring diffuser. The proposed design, as shown in Figure 2.28(a), involves two horizontal discs (termed the primary and secondary splitters) enclosed by a hemispherical guide. Incoming fluid from the top impinges vertically on the splitter first and proceeds either upwards or downwards depending on the temperature.

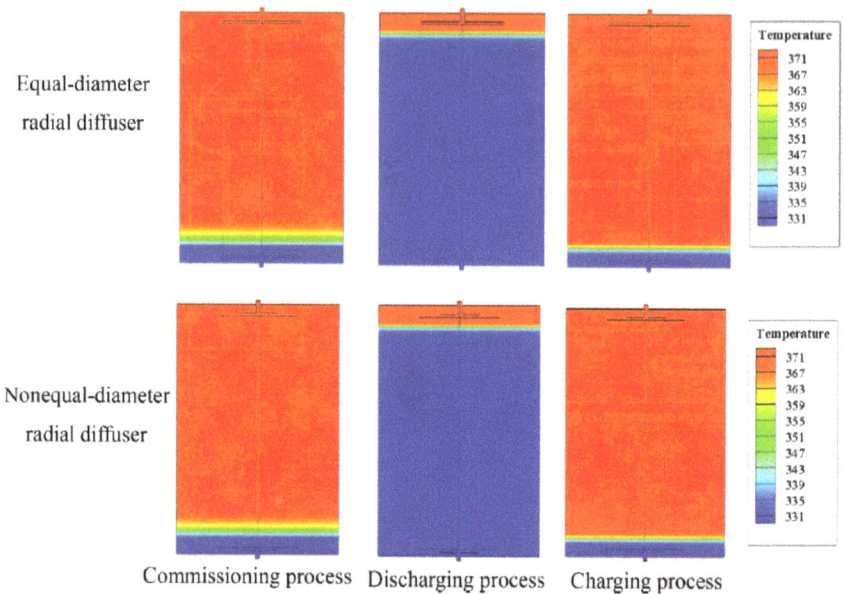

Figure 2.27.   Deng *et al.* found negligible difference in stratification performance between equal and unequal plate diameter diffusers [33].

The mechanism allows the inertial jet and recirculation flows to be confined within the diffuser, as portrayed in Figure 2.28(b).

Key findings have highlighted the impact of the gravitational current length and the induced vortices on the hot-and-cold fluid thermal blending. The proposed hemispherical diffuser has been observed to mitigate this effect considerably at an acceptable pressure drop of around 700 Pa at 27 L/min flow rate. Comparing the numerical solutions revealed that the conventional diffusers recorded depreciating stratification at higher flow rates, while the opposite is true for the proposed diffuser, which recorded a 6% improvement at 27 L/min in contrast to other diffusers. This is depicted in Figure 2.28(c), which illustrates the relative differences between the mixing occurring in a direct entry inlet and the proposed diffuser design.

2.3.   Commercial and Industrial Examples

TES systems play a central role in the design and implementation of centralized thermal distribution networks such as district heating and

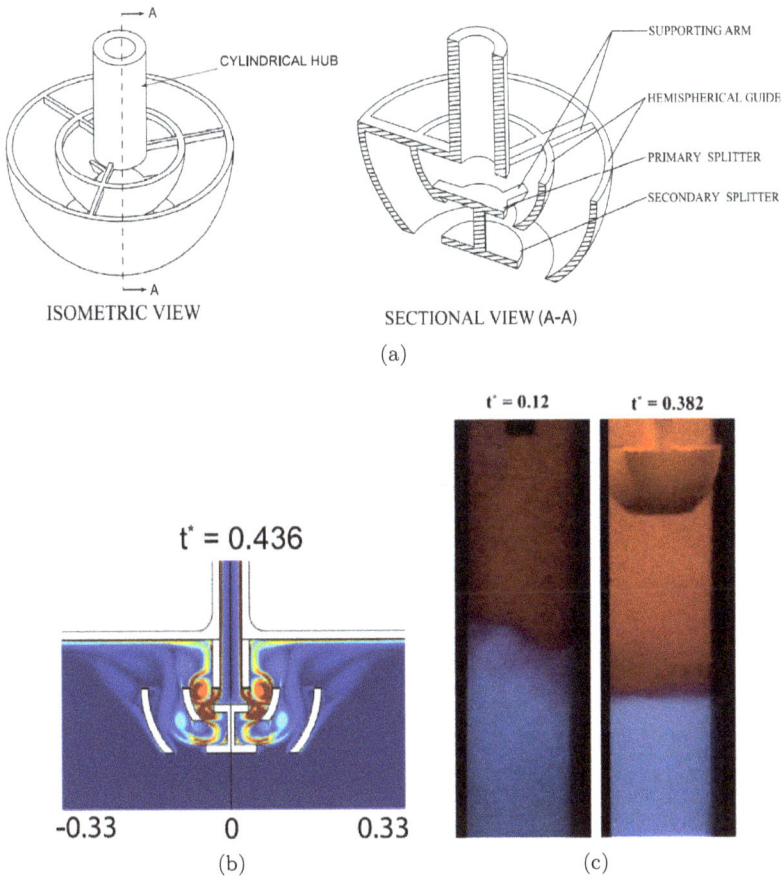

Figure 2.28. (a) Design and internal structure of hemispherical diffuser; (b) vortices induced during operation within the diffuser; and (c) comparison between direct fluid inlet and hemispherical diffuser for 27 L/min inlet flow rate [34].

cooling systems. District cooling systems (DCS), in particular, have been deployed in many tropical climate contexts, where urban cooling demand constitutes a significant portion of the annual urban energy consumption. Cooling accounts for 24% of household energy demand and 40–50% of commercial energy demand [35]. TES systems are instrumental in DCS operations as they facilitate thermal balancing during the time differences between daily peak cooling demand and supply. Furthermore, the differences in peak and off-peak electrical tariffs yield significant operational cost savings. TES are charged

A District Cooling System

Imagine a giant air-conditioner that can cool an entire district of buildings, rather than just individual buildings – but greener and more energy efficient. How does it work?

1. Chilled water is generated in a central cooling plant.

2. A closed loop network of underground insulated pipes distributes the chilled water to each customer's building.

3. When the chilled water reaches the customer's building, energy transfer stations within each building circulate the cold energy from the network into the building's air-conditioning system, which dehumidifies and cools the air.

4. The warmer water is then circulated to the cooling plant, via the return pipes, to be chilled again. The whole process repeats itself.

5. Thermal storage tanks (if used*), are designed to store cold energy, in the form of ice or chilled water. Thermal storage tanks help to regulate cooling demand and provide resilience.

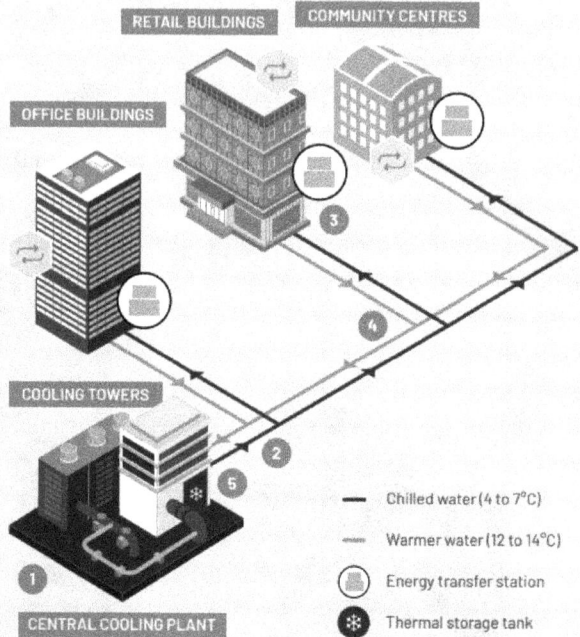

RETAIL BUILDINGS   COMMUNITY CENTRES

OFFICE BUILDINGS

COOLING TOWERS

CENTRAL COOLING PLANT

— Chilled water (4 to 7°C)

⁓ Warmer water (12 to 14°C)

⊛ Energy transfer station

❄ Thermal storage tank

*Not all district cooling system plants deploy thermal storage tanks.

Figure 2.29.   Illustration of a DCS by SP Group Pte Ltd [35].

during off-peak periods at a lower operating cost and discharged to meet the active cooling demand during peak hours, ensuring a consistent chiller load in order to maximize the chiller's operating efficiency, as shown in Figure 2.29.

In Singapore, key industrial players in the DCS scene include Singapore Power Pte Ltd (SP Group) and Keppel District Heating and Cooling Systems (KDHCS). SP Group operates key district cooling facilities in Marina Bay Sands, Tengah, and Tampines Eco Town [35], while KDHCS has existing facilities in high-profile commercial districts that include Changi Business Park, *Biopolis* and *Mediapolis* at the One-North district, and the Woodlands Wafer

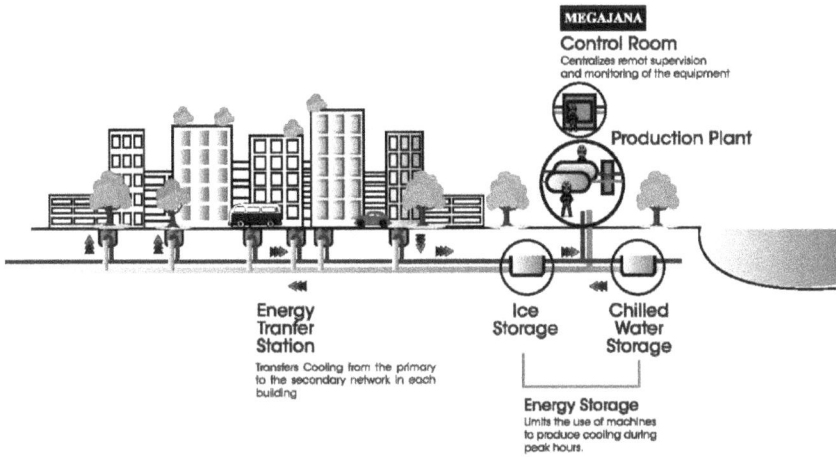

Figure 2.30.   Schematic representation of the DCS at Cyberjaya by Pendinginan Megajana Sdn Bhd [38].

Fab Park [36]. *Engie*, which operates in multiple southeast Asian markets has also undertaken joint ventures with local district cooling providers in several ASEAN locations such as Malaysia and the Philippines. For instance, the Cyberjaya District Cooling System in Selangor, Malaysia, operated by Pendinginan Megajana Sdn Bhd, boasts a cooling capacity of up to 22,000 tonnes of refrigeration (RT) with over 97,500 refrigeration tonne-hours of TES capacity; servicing more than 51 buildings over 3 km^2 of land area [37,38]. As illustrated in Figure 2.30, the system features an ice-based latent TES working in tandem with a chilled water-based STES to support the combined cooling load operating between 6°C and 13°C.

As centralized thermal distribution networks become more prevalent and integrated into our energy landscape, so will the application of TES systems and the evolution of innovative solutions arising from academic research.

2.4.   Conclusion

STES systems constitute the most fundamental and primitive branch of TES systems. However, there is still an abundance of research opportunities available to improve and optimize its design and application. In this chapter, the fundamentals of STES systems ranging from the performance indices to common characterization methods, both semi-empirical and numerical, are covered. Key stratification concepts are also illustrated with novel research from recent times to demonstrate the extent of innovation in today's research studies on fundamental STES systems. Lastly, key industrial players and the associated applications of TES in districtwide thermal management systems are introduced to illustrate the importance of STES technology in achieving sustainable large-scale cooling.

References

[1] E. González-Roubaud, D. Pérez-Osorio and C. Prieto, Review of commercial thermal energy storage in concentrated solar power plants: Steam vs. molten salts, *Renewable and Sustainable Energy Reviews*, vol. 80, pp. 133–148, 2017.

[2] E. Oró, A. Castell, J. Chiu, V. Martin and L. F. Cabeza, Stratification analysis in packed bed thermal energy storage systems, *Applied Energy*, vol. 109, pp. 476–487, 2013.

[3] M. A. A. Majid, M. Muhammad, C. C. Hampo and A. Bt Akmar, Analysis of a thermal energy storage tank in a large district cooling system: A case study, *Processes*, vol. 8, no. 9, p. 1158, 2020.

[4] M. A. Abd Maji and K. K. Looi, Evaluations of thermocline and half cycle figure of merit of a thermal energy storage tank, in *MATEC Web of Conferences*, 2017.

[5] B. Koçak, A. I. Fernandez and H. Paksoy, Review on sensible thermal energy storage for industrial solar applications and sustainability aspects, *Solar Energy*, vol. 209, pp. 135–169, 2020.

[6] Z. Lavan and J. Thompson, Experimental study of thermally stratified hot water tanks, *Solar Energy*, vol. 19, pp. 519–524, 1977.

[7] A. Castell, M. Medrano, C. Sole and L. Cabeza, Dimensionless numbers used to characterize stratification in water tanks for discharging at low flow rates, *Renewable Energy*, vol. 35, pp. 2192–2199, 2010.

[8] J. D. Chung, S. H. Cho, C. S. Tae and H. Yoo, The effect of diffuser configuration on thermal stratification in a rectangular storage tank, *Renewable Energy*, vol. 33, no. 10, pp. 2236–2245, 2008.

[9] Y. P. Chandra and T. Matsuka, Stratification analysis of domestic hot water storage tanks: A comprehensive review, *Energy & Buildings*, vol. 187, pp. 110–131, 2019.

[10] N. Altuntop, M. Arslan, V. Ozceyhan and M. Kanoglu, Effect of obstacles on thermal stratification in hot water storage tanks, *Applied Thermal Engineering*, vol. 25, no. 14–15, pp. 2285–2298, 2005.

[11] E. Andersen, S. Furbo and J. Fan, Multilayer fabric stratification pipes for solar tanks, *Solar Energy*, vol. 81, no. 10, pp. 1219–1226, 2007.

[12] D. Mangold and L. Deschaintre, *Seasonal Thermal Energy Storage*, International Energy Agency, Stuttgart, 2016.

[13] S. Furbo, Using water for heat storage in thermal energy storage (TES) systems, in *Advances in Thermal Energy Storage Systems — Methods and Applications*, Cambridge, UK: Elsevier, 2015, pp. 31–45.

[14] S. E.-D. Fertahi, A. Jamil and A. Benbassou, Review on Solar Thermal Stratified Storage Tanks (STSST): Insight on stratification studies and efficiency indicators, *Solar Energy*, vol. 176, pp. 126–145, 2018.

[15] J. Davidson, D. Adams and J. Miller, A coefficient to characterize mixing in solar water storage tanks, *Journal of Solar Energy Engineering*, vol. 116, no. 2, pp. 94–99, 1994.

[16] J. Rendall, A. Abu-Heiba, K. Gluesenkamp, K. Nawaz, W. Worek and A. Elatar, Nondimensional convection numbers modeling thermally stratified storage tanks: Richardson's number and hot-water tanks, *Renewable and Sustainable Energy Reviews*, vol. 150, p. 111472, 2021.

[17] U. Jordan and S. Furbo, Thermal stratification in small solar domestic storage tanks caused by draw-offs, *Solar Energy*, vol. 78, no. 2, pp. 291–300, 2005.

[18] J. Fernandez-Seara, F. J. Uhía and J. Sieres, Experimental analysis of a domestic hot water storage tank. Part II: Dynamic mode of operation, *Applied Thermal Engineering*, vol. 27, pp. 37–44, 2007.

[19] V. Palombo and A. Frazzica, Application of numerical methods for the design of thermocline thermal energy storage: Literature review and critical analysis, *Journal of Energy Storage*, vol. 46, p. 103875, 2022.

[20] H. Khurana, R. Majumdar and S. K. Saha, Response surface methodology-based prediction model for working fluid temperature during stand-alone operation of vertical cylindrical thermal energy storage tank, *Renewable Energy*, vol. 188, pp. 619–636, 2022.

[21] K. Ismail and R. Stuginsky, Parametric study on possible fixed bed models for PCM and sensible heat storage, *Applied Thermal Engineering*, vol. 19, pp. 757–788, 1999.

[22] S. Ergun, Fluid flow through packed columns, *Chemical Engineering Progress*, vol. 48, no. 2, pp. 89–94, 1952.

[23] J. Beek, Design of packed catalytic reactors, *Advanced Chemical Engineering*, vol. 3, pp. 203–271, 1962.

[24] N. Wakao, S. Kaguei and T. Funazkri, Effect of fluid dispersion coefficient on particle-to-fluid heat transfer coefficients in packed beds, *Chemical Engineering Science*, vol. 34, pp. 325–336, 1979.

[25] T. Galloway and B. Sage, A model of the mechanism of transport in packed, distended and fluidized beds, *Chemical Engineering Science*, vol. 25, pp. 495–516, 1970.

[26] J. Bédécarrats, J. Castaing-Lasvignottes, F. Strub and J. Dumas, Study of a phase change energy storage using spherical capsules. Part II: Numerical modelling, *Energy Conversion and Management*, vol. 50, no. 10, pp. 2537–2546, 2009.

[27] M. Sozen and K. Vafai, Analysis of energy and momentum transport for fluid flow through a porous bed, *Journal of Heat Transfer*, vol. 112, no. 3, pp. 690–699, 1990.

[28] B. Baeten, T. Confrey, S. Pecceu, F. Rogiers and L. Halsen, A validated model for mixing and buoyancy in stratified hot water storage tanks for use in building energy simulations, *Applied Energy*, vol. 172, pp. 217–229, 2016.

[29] A. Rahman and A. D. Smith, Predicting heating demand and sizing a stratified thermal storage tank using deep learning algorithms, *Applied Energy*, vol. 228, pp. 108–121, 2018.

[30] Z. Yang, H. Chen, L. Wang, Y. Sheng and Y. Wang, Comparative study of the influences of different water tank shapes on thermal energy storage capacity and thermal stratification, *Renewable Energy*, vol. 85, pp. 31–44, 2016.

[31] Q. Sun, H. Wang, K. Dong, L. Tv and Z. Kang, Experimental study on the thermal performance of a novel physically separated chilled water storage tank, *Journal of Energy Storage*, vol. 40, p. 102628, 2021.

[32] L. Kocijel, V. Mrzljak and V. Glažar, Pressure drop in large volumetric heat storage tank radial plate diffuser, *Journal of Energy Storage*, vol. 29, p. 101350, 2020.

[33] Y. Deng, D. Sun, M. Niu, B. Yu and R. Bian, Performance assessment of a novel diffuser for stratified thermal energy storage tanks — The nonequal-diameter radial diffuser, *Journal of Energy Storage*, vol. 35, p. 102276, 2021.

[34] D. R. Parida, S. Advaith, N. Dani and S. Basu, Assessing the impact of a novel hemispherical diffuser on a single-tank sensible thermal energy storage system, *Renewable Energy*, vol. 183, pp. 202–218, 2022.

[35] Singapore Power Pte Ltd, Taking the heat off cooling: A greener way to cool, Singapore Power Pte Ltd, Singapore, 2021.

[36] Keppel Corporation Pte Ltd, Singapore plants [Online]. Available: https://www.keppeldhcs.com/singapore_plants.html.

[37] ENGIE, District cooling providing energy to communities through efficient and decarbonised district energy solutions, 2021 [Online]. Available: https://www.engie-sea.com/district-cooling.

[38] Pendinginan Megajana Sdn. Bhd., District cooling, 2021 [Online]. Available: http://megajana.com.my/megajana/district-cooling/.

Chapter 3

Latent Thermal Energy Storage Systems

Abstract

Latent thermal energy storage (LTES) systems utilize the phase-change properties of the storage media to allow high thermal energy densities to be contained within narrower temperature bands. Latent heat is released or absorbed during a material's phase-change process and this heat is many times higher than its specific heat capacity while occurring at a near-constant temperature. This allows for higher thermal energy densities for applications in contrast to sensible TES systems. In the ensuing sections of this chapter, the fundamentals of PCMs and the thermal transfer of LTES systems will be first covered. Second, the various classifications and manifestations of LTES systems will be introduced. Finally, several real-life examples of LTES applications will be described to illustrate the increasing impact that this form of TES has on the dynamic energy distribution landscape.

Keywords: Phase-change material; differential scanning calorimetry; microencapsulation; packed-bed systems.

3.1. Fundamentals of Phase-Change Materials

The phase-change material (PCM) forms the fundamental core of an Latent thermal energy storage (LTES) system. In contrast to sensible TES systems, the PCM allows an LTES to achieve higher levels of thermal energy density over smaller operating temperature bounds via the phase-change process. In principle, besides the traditional solidification/melting process, it is extended to all phase-change processes, including evaporation/condensation and solid–solid phase-change processes. The former, however, is largely unpractised outside of liquid air energy storage (LAES) applications, which have the primary purpose of re-gasification for power generation. The latter case of solid–solid PCMs has received increasing attention in recent years, but unfortunately it is still largely confined to studies conducted at an experimental scale. This section will specifically focus on solid–liquid PCMs.

3.1.1. Classification of PCM types

Similar to the STES systems, the storage material is primarily responsible for determining the operating temperature range and storage capacity of the LTES. PCMs, although existing in a wide range of forms and states, are often limited to operating within the nominal phase-change temperature for that material class. Li [1] has comprehensively summarized the varying types of solid–liquid PCMs

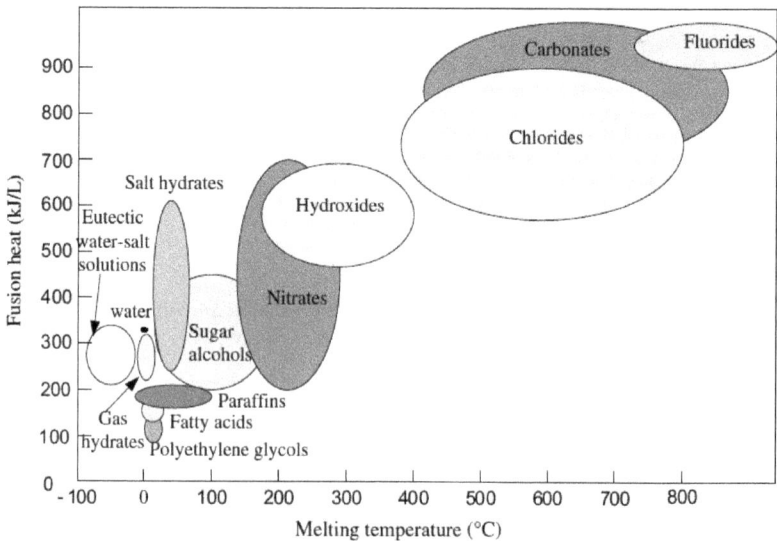

Figure 3.1. Distribution of solid–liquid PCMs over the various phase-change temperatures and the corresponding latent heat of fusion [1].

over the various phase-change temperatures and their corresponding latent heat of fusion as shown in Figure 3.1.

Inorganic PCMs, like eutectic salt-water solutions, primarily occupy phase-change temperatures at cryogenic ranges of $-100°C$ to $0°C$ but occupy moderate latent heats spanning 200–300 kJ/L. Conventional urban cooling temperature range of 0–15°C necessitates organic PCMs like paraffins and fatty acids to realize lower latent heat capacities of around 100–200 kJ/L. Paraffin and salt hydrates, operating at higher temperatures ranging from 30°C to 100°C, tend to be more prevalent. This temperature range is conventional for urban space heating applications, especially in temperate or low-temperature climates with high heating demand. For temperatures in excess of 100°C up to as high as 800–900°C, molten salts such as chlorides, fluorides, and carbonates are often employed.

Selection of the appropriate PCM for a specific application is strongly dependent on its thermophysical properties, most significantly its phase-change temperature. In the case of commercially produced PCMs, this is often available from the manufacturer's specifications, but with more generalized mixtures such as eutectic-salt solutions, the binary phase diagram is often employed as a means of quickly identifying its approximate properties. This can be seen, for instance, via the binary phase diagram for NaCl–H$_2$O at 100 kPa as shown in Figure 3.2. Point E at the intersection of the 23.3 wt% NaCl and $-21.1°C$ indicates the eutectic point of the solution.

3.1.2. *PCM characterization methods*

To characterize other types of PCMs, however, more in-depth investigative methods are often required. The two most commonly used methods for PCM characterization are differential-scanning calorimetry (DSC) and T-history methods [3]. In the DSC method, the sample and a reference sample are subject to a transient temperature profile, while their relative heat flows are measured [4] as depicted in Figure 3.4(a). The T-history method, on the other hand, takes a sample and its reference (usually a known material) and subjects them to ambient conditions from an initial condition as illustrated in Figure 3.3.

Their temperature profiles based on progressive cooling or heating to ambient temperatures are then used to derive the sample

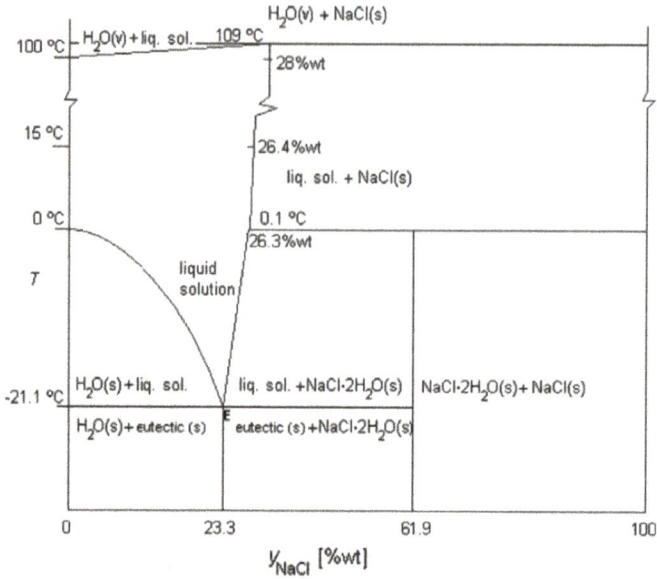

Figure 3.2.   Binary phase diagram for NaCl–H$_2$O at 100 kPa [2]. Point E at the intersection of the 23.3 wt% NaCl and –21.1°C indicates the eutectic point of the solution.

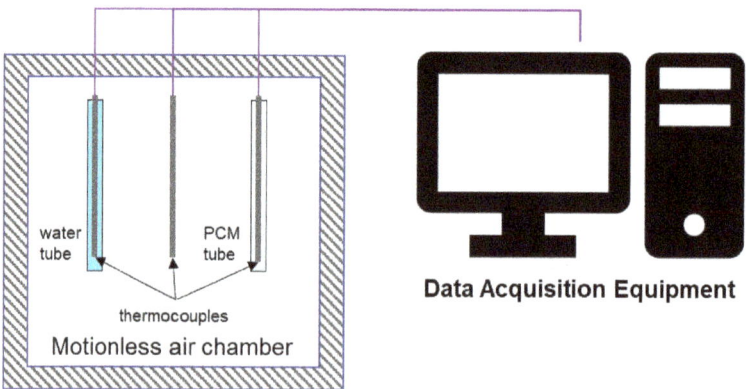

Figure 3.3.   A T-history test setup with the PCM and its reference samples within a chamber [5].

material's heat capacitance [5]. Figure 3.4(b) demonstrates the T-history temperature profiles relative to the ambient conditions the samples are subject to. Rathgeber *et al.* [3] argue that the key differences between these methods are the sample size and temperature

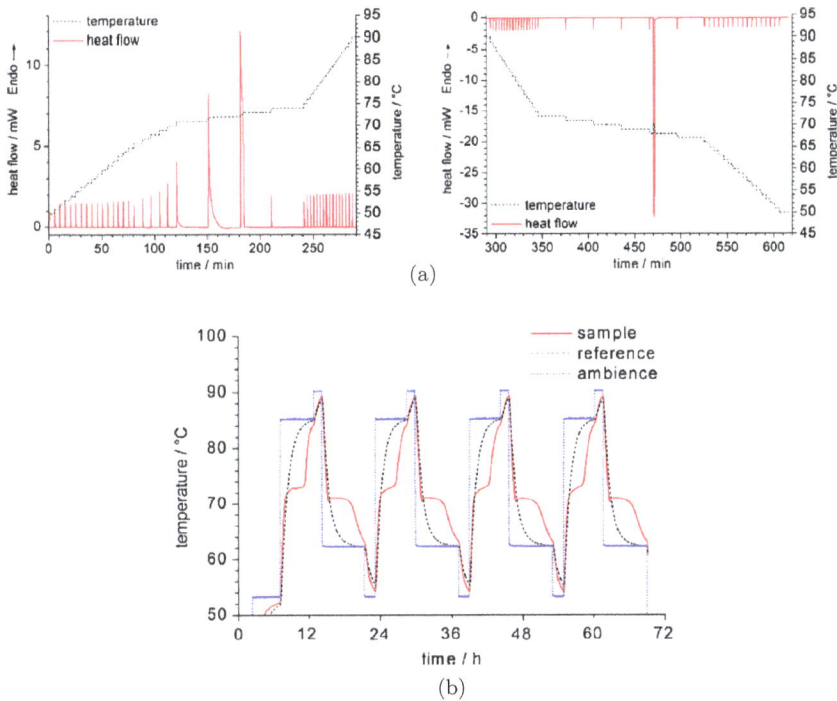

Figure 3.4.   (a) A heating run (left) and a cooling run (right) for DSC testing of a PCM sample. The heat flow lines indicate the enthalpy increase or decrease of the sample, while the dotted lines indicate the temperature profile that the sample is subjected to. (b) The temperature–time curves of a T-history method. The blue lines indicate the ambient conditions that the sample and reference are subjected to, while the red and dotted lines indicate the transient temperature profiles of the sample and reference, respectively [3].

profiles imposed on the samples. DSC methods often require very small sample sizes to reduce the impacts of higher thermal capacitances on the response, while T-history methods tend to use up to 1,000 times the sample size.

The scanning rate is a key parameter during a DSC test which determines the rate of temperature change that the sample is subjected to throughout the test. As higher rates of cooling are known to induce supercooling effects in PCMs, regulating the scan rates reveals not only the usual thermophysical properties but also the susceptibility of the PCM to supercooling, as can be seen in Figure 3.5, through a DSC test conducted on a paraffin mixture containing n-tetradecane and n-hexadecane.

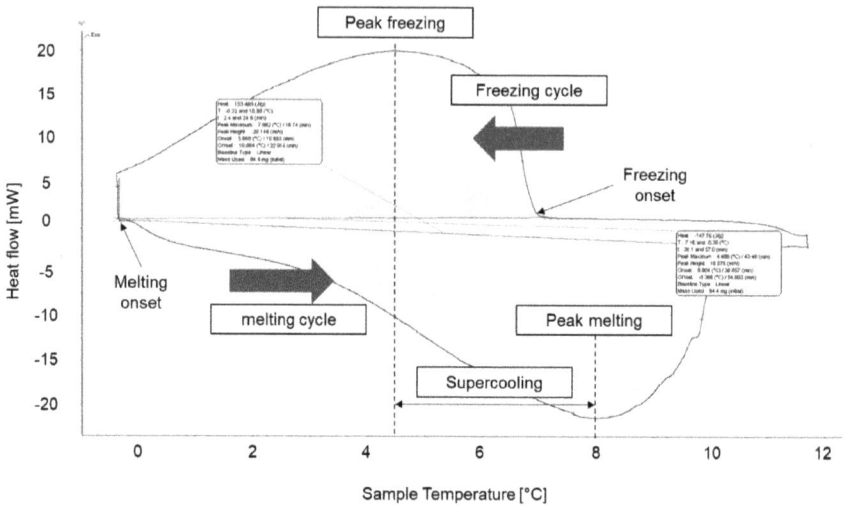

Figure 3.5. DSC test of a paraffin mixture with 50% n-tetradecane and 50% n-hexadecane at a scan rate of 0.5 K/min. The temperature differences between the peak melt and freeze temperatures indicate the degree of supercooling at this rate of cooling. The apparatus used was a Setaram Sensys Evo DSC system [6].

The DSC highlights the heat flow to and from the sample (relative to the reference) while being subjected to a temperature profile between 0°C and 12°C. The top segment reflects the freezing (cooling) cycle, while the bottom segment indicates the melting (heating) cycle. As the sample is first cooled from 12°C to 0°C, the peak heat flow is observed at 4.5°C, therefore, the peak freeze point during this stage is considered to be 4.5°C. As the PCM is heated from 0°C to 12°C, the peak heat flow is observed to be at 7.96°C; demarcating the peak melting point of the PCM. This temperature difference indicates that the degree of supercooling, experienced by the PCM at 0.5 K/min scan rate, is around 2.5°C. The temperature width of the freezing and melting process further reflects the overall phase-change process temperature boundary, which is located between 0°C and 10°C. This information is essential not just during the PCM selection process but also in modelling and analyzing LTES systems, where accounting for supercooling and corresponding phase-change temperature range are crucial in predicting the transient thermal performance of the LTES. Karthikeyan and Velraj [7] classified the melting process of a PCM into three stages:

(1) Sensible heat transfer in solid phase;
(2) The phase-change process involves both sensible and latent heat transfer, depending on the instantaneous solid/liquid fraction of the PCM; and
(3) Sensible heat transfer occurs in the liquid phase after complete melting.

The effective specific heat and thermal conductivity properties of the PCM during each of these phases can then be described accordingly as

$$
\begin{cases}
c_p\left(T\right) = c_{pl} & T > T_{pl} & \text{Liquid} \\
c_p\left(T\right) = \frac{h_L}{T_{pl}-T_{ps}} + \lambda \cdot c_{ps} + (1-\lambda) \cdot c_{pl} & T_{ps} \leq T \leq T_{pl} & \text{Phase-change} \\
c_p\left(T\right) = c_{ps} & T < T_{ps} & \text{Solid}
\end{cases}
$$

$$(3.1)$$

$$
\begin{cases}
k_p\left(T\right) = k_{pl} & T > T_{pl} & \text{Liquid} \\
k_p\left(T\right) = \lambda \cdot k_{ps} + (1-\lambda) \cdot k_{pl} & T_{ps} \leq T \leq T_{pl} & \text{Phase-change} \\
k_p\left(T\right) = k_{ps} & T < T_{ps} & \text{Solid}
\end{cases}
\qquad (3.2)
$$

where c_p(kJ/kg·K), k (W/mK), λ and h_L (kJ/kg) refer to the specific heat thermal conductivity, solid fraction, and latent heat of fusion of the PCM, respectively. The subscripts l and s refer to the liquid and solid phases of the PCM, respectively. $T(°C)$ refers to the temperature of the PCM, while T_{ps} and $T_{pl}(°C)$ refer to the lower and upper bounds of the PCM phase-change region.

3.2. Fundamentals of LTES Systems

While the fundamental behaviour of an LTES is largely governed by the PCM and its phase-change behaviour, simply solving for the phase-change process itself is insufficient to predict the LTES performance. This is attributed to the complex time-varying heat transfer and thermal stratification involved, which has been covered in significant detail in Chapter 2. The phase-change process is described through the Stefan problem, which broadly covers models describing thermodiffusive behaviours accompanied by phase transformations of the media and, more specifically, a moving phase-separation boundary [8]. The Stefan problem has been widely covered in the literature and, considering the diversity that involves the tedious analytical approaches to solve it, going into extensive details is beyond the scope of this chapter.

Groulx [9] opined that models that can directly compare the predicted performance of LTES based on design and operating parameters are few. Many LTES works have relied primarily on numerical methods validated with experimental data. However, many of these computational methods tend to incur exorbitant computational power and cost. Many model simplifications have been made to mitigate this issue but often lead to poor accuracy or poor generalizability. This section covers the basic principles that govern various manifestations of LTES systems (non-encapsulated vs encapsulated).

3.2.1.   *Non-encapsulated LTES systems*

Non-encapsulated LTES systems constitute the most common and basic form of PCM application in LTES. By having the PCM contained in the tank with a heat transfer fluid (HTF) running through the system yields maximum volume fraction of the PCM within the LTES. For industrial applications, these often take the form of ice-on-coil storage systems such as portrayed in Figure 3.6. Ice-on-coil systems feature cooling coils carrying a sub-zero coolant such as aqueous glycol solution flowing through a water tank. The water freezes on the exterior of the cooling coils while a small remaining unfrozen fraction of the TES acts as the HTF during discharge. In other manifestations of this configuration, all the water freezes and a separate set of tubes carrying the HTF is also present.

Non-encapsulated systems, such as ice-on-coil systems, are easier to model mathematically, especially in cases where stratification is not a priority. Heat transfer between the PCM and HTF is accounted for via tube-side flow. Tay *et al.* [11,12] proposed an effectiveness–NTU method; similar to the method employed to design conventional heat exchangers.

Key assumptions of the approach are as follows:

(1) Constant inlet HTF velocity and temperature;
(2) LTES outer walls are assumed adiabatic;
(3) Initial conditions are uniform throughout LTES. PCM is initially completely solid for a melting process and completely fluid for a freezing process;

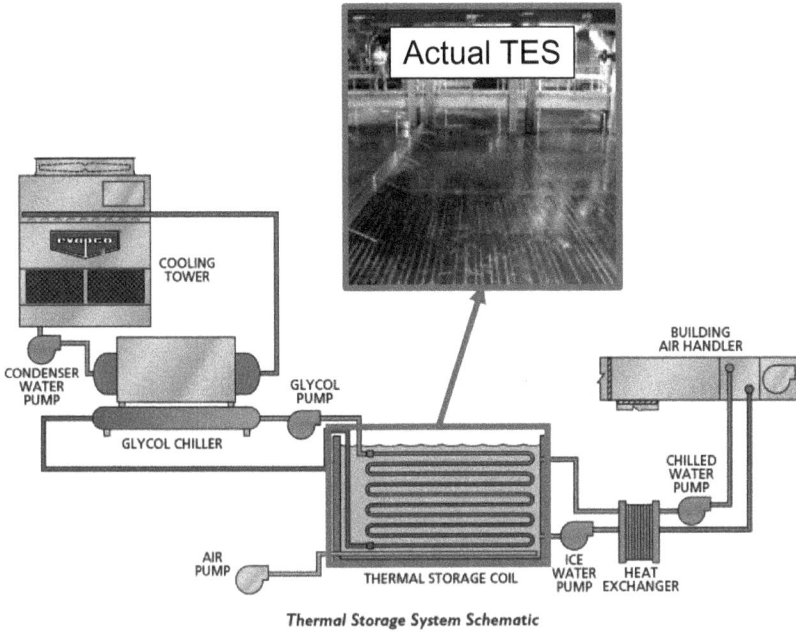

Figure 3.6. An ice-on-coil TES by EvapCo [10].

(4) Model is axisymmetric;
(5) Constant thermophysical properties for tubes, HTF, and PCM; and
(6) Natural convection in the liquid PCM is deemed negligible.

The effectiveness–NTU method entails first defining the effectiveness, ε, of the LTES as the ratio of the temperature difference across the inlet and outlet to the temperature difference between the inlet temperature and the PCM phase-change temperature,

$$\varepsilon = \frac{T_{\text{in}} - T_{\text{out}}}{T_{\text{in}} - T_{\text{PC}}} \tag{3.3}$$

$$\varepsilon = 1 - \exp\left(-\text{NTU}\right) \tag{3.4}$$

where T_{in}, T_{out}, and $T_{\text{PC}}(^{\circ}\text{C})$ refer to the inlet, outlet, and phase-change temperatures of the LTES, respectively. Equation (3.4) describes the relationship between this effectiveness and the number of heat transfer units (NTU) wherein the value can be derived empirically via prior experimental data [12], thereby leading to the relationship as follows:

$$\varepsilon = 1 - \exp\left(-0.0199\left(\frac{A}{\dot{m}}\right)\right) \tag{3.5}$$

where A (m^2) and $\dot{m}$(kg/s) refer to the heat transfer area and mass flow rate of the HTF through the LTES, respectively. The NTU can then be described as

$$\text{NTU} = \frac{UA}{\dot{m}c_p} \tag{3.6}$$

where U refers to the overall heat transfer coefficient described by

$$U = \frac{1}{R_T} = \frac{1}{R_{\text{HTF}} + R_{\text{tube}} + R_{\text{PCM}}}$$

$$= \left(\frac{1}{2\pi r_i l h_f} + \frac{\ln\left(\frac{r_o}{r_i}\right)}{2\pi k_{\text{tube}} l} + \frac{\ln\left(\frac{r_o}{r_i}\right)}{2\pi k_{\text{PCM}} l}\right) \tag{3.7}$$

$$h_f = \frac{\text{Nu} \cdot k_f}{2r_i} \tag{3.8}$$

$$\text{Nu} = 3.66 + \frac{\left(0.0668\left(\frac{2r_i}{l}\right)\text{RePr}\right)}{1 + 0.04\left(\left(\frac{2r_i}{l}\right)\text{RePr}\right)^{\frac{2}{3}}} \quad \text{For laminar flow}$$

$$\text{Nu} = 0.023\text{Re}^{0.8}\text{Pr}^n \tag{3.9}$$

$(n = 0.3$ for cooling and 0.4 for heating)   For turbulent flow

$$\text{Re} = \frac{\dot{m}_{\text{HTF}}(2r_i)}{A_c\mu} \tag{3.10}$$

where R_T (mK/W) refers to the overall thermal resistance. R_{HTF}, R_{tube}, and R_{PCM} refer to the respective thermal resistances of the HTF, tube, and PCM, respectively, as depicted in Figure 3.7. The right-hand side of Equation (3.7) represents the various thermal resistances. r_i and r_o (m) refer to the inner and outer diameter of the tubes, respectively, and l (m) refers to the length of the tube. k_{tube} and k_{pcm}(W/mK) refer to the thermal conductivities of the tube and PCM, respectively. Nu and Re, described by Equations (3.6) and (3.7), refer to the Nusselt and Reynolds number of the

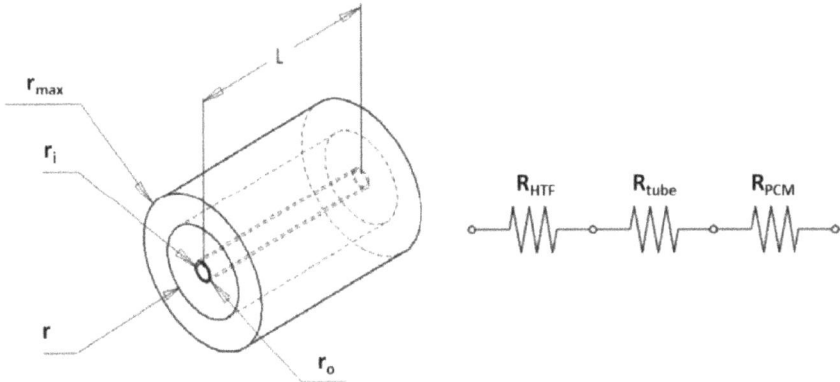

Figure 3.7.   Thermal resistance circuit for the HTF tube [11].

HTF flow, respectively, while the values A_C (m^2) and μ (Pa $\cdot$ S) refer to the cross-sectional area of the tube and dynamic viscosity of the HTF, respectively.

Accounting for the PCM solid fraction, the overall average effectiveness is then summarized as

$$\bar{\varepsilon} = \int_0^1 1 - \exp\left(-\frac{1}{\dot{m}_{\mathrm{HTF}} c_p R_T}\right) d\lambda \qquad (3.11)$$

This expression cannot be determined explicitly and needs to be numerically computed. It is worth noting that when natural convection is insignificant within the PCM in the liquid phase, only conduction is considered in its formulation.

The available numerical approaches to simulate and study LTES are largely similar to those described in Chapter 2. It is noteworthy that these numerical approaches tend to revolve around making simplifications to reduce computational power and time, as highlighted in Figure 3.8.

3.2.2.   *Encapsulated LTES systems*

Encapsulated systems, in contrast, have been receiving greater attention in recent years due to their potential to promote heat transfer between PCM and HTF, as well as offering better performance consistency. Encapsulating and interfacing methods for PCMs in

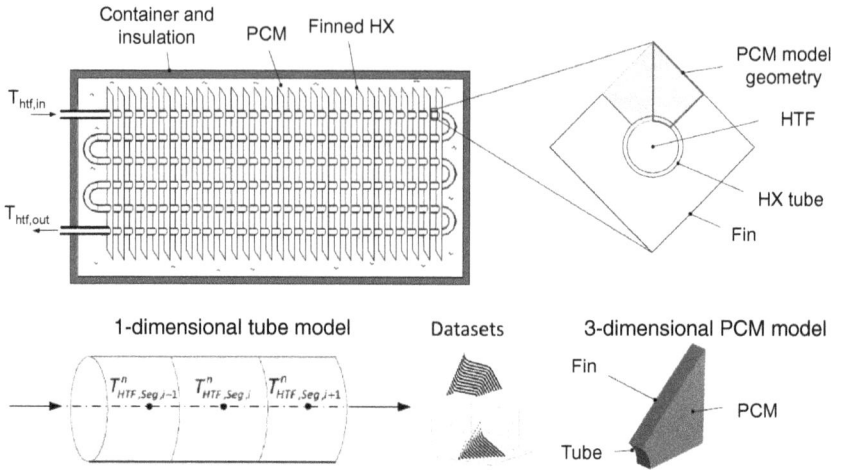

Figure 3.8. Numerical modelling of a finned-tube LTES (with PCM on shell side) by Waser *et al.* [13], showing the simplification of a complex structure into a 1D tube model and a reduced 3D PCM model.

LTES systems take the form of macroencapsulation or micro-/nano-encapsulation. Macroencapsulation involves PCM being contained in shells that take the physical forms of spheres, tubes, slabs, or other novel variations of these shapes, often deployed en masse in storage tanks. Macroencapsulating methods are more widely prac-tised in industry owing to simplicity of application. Micro- or nano-encapsulation involves some level of chemical or physicochemical treatment as elaborated in Chapter 1. The PCM is contained in a shell at micro- or nano-scales via novel methods such as spray dry-ing, polymerization, or sol–gel methods, as shown in Figure 3.9(a). Research on this form of encapsulation is still in its infancy but has been gaining greater traction in recent times.

One non-conventional PCM/HTF interfacing involves the use of organic/inorganic frameworks. These are often also known as form- or shape-stabilized PCMs (SPCM). They employ the frameworks to maintain the "solid" form of the PCM even at temperatures above its melting point. The frameworks used are porous polymer-based materials, and the PCMs are "encapsulated" onto the pores or poly-mer chains of the materials, as portrayed in Figure 3.9(b). Thus far, research has been confined largely to lab-scale studies.

(a) (b)

Figure 3.9. (a) n-Octacosane microcapsules under scanning electron microscopy (SEM) imaging and (b) schematic representation of shape-stabilized PCM in a porous inorganic framework. The porous material constitutes a form of "encapsulation" for the PCM even in its liquid phase allowing it to be form-stabilized under all states [14].

While micro-/nano-encapsulated PCMs and shape-stabilized PCMs are likely to represent the next frontier of PCM and LTES technological development, the following section will focus on more conventional macroencapsulated PCMs, which have greater industrial applicability.

3.2.2.1. *Packed-bed LTES*

Currently, packed-bed LTES systems form the bulk of encapsulated LTES studies found in the literature. By encapsulating PCM in spheres or sphere-adjacent shapes, a uniform and high specific heat transfer surface area is achieved when packed into a storage tank. The amount of spheres or encapsulated PCMs that can be packed into a storage tank is, however, constrained by a theoretical limit which is dependent on the relative diameter ratios. Table 3.1 details seven different correlations documented in the literature that predict the volume fraction of equal-sized spherical particles in a cylindrical column based on the relative diameter ratio. It is apparent that the bed packing density approaches a theoretical maximum of approximately 0.62 when the relative diameter D/d exceeds 10. This is illustrated in Figure 3.10, where it is demonstrated that the packing density plateaus as D/d approaches 100 and beyond.

Table 3.1. Computation of bed porosity (packing density) by various authors. The diameter ratio refers to the ratio of the packed bed diameter to the particle diameter.

Author/Reference	Formulation	Validity	Diameter ratio					
			2	5	10	20	50	100
Sato et al. [15]	$1-\left(0.3494+\frac{0.4381}{\frac{D}{d}}\right)$	$\frac{D}{d}>2.5$	—	0.563	0.606	0.629	0.642	0.646
Dixon [16]	$1-\left(0.4+\frac{0.05}{\frac{D}{d}}+\frac{0.412}{\left(\frac{D}{d}\right)^2}\right)$	$\frac{D}{d}>2$	0.472	0.574	0.591	0.597	0.599	0.600
Fand and Thinakaran [17]	$1-\left(\frac{0.151}{\frac{D}{d}-1}+0.36\right)$	$\frac{D}{d}>2.033$	0.489	0.602	0.623	0.632	0.637	0.639
Foumeny et al. [18]	$1-\frac{0.383+0.254\left(\frac{D}{d}\right)^{-0.923}}{\sqrt{0.723\frac{D}{d}-1}}$	$\frac{D}{d}>1.866$	0.416	0.581	0.605	0.613	0.616	0.616
Zou and Yu [19]	$1-\left(0.372+0.002\left(\exp\left(\frac{15.306}{\frac{D}{d}}\right)-1\right)\right)$	$\frac{D}{d}>3.95$	—	0.587	0.621	0.626	0.627	0.628
Benyahia and O'Neill [20]	$1-\left(0.39+\frac{1.74}{\left(\frac{D}{d}+1.14\right)^2}\right)$	$1.5<\frac{D}{d}<50$	0.433	0.563	0.597	0.606	0.609	0.610
Ribeiro et al. [21]	$1-\left(0.373+0.917\exp\left(-0.824\frac{D}{d}\right)\right)$	$2<\frac{D}{d}<19$	0.450	0.612	0.627	0.627	0.627	0.627

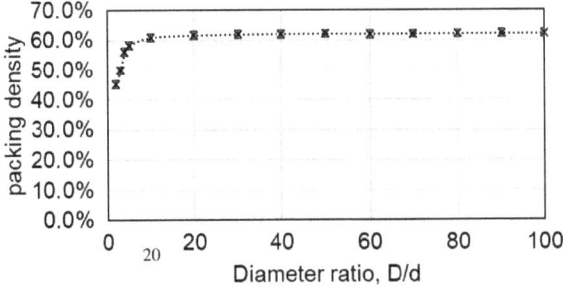

Figure 3.10.   Prediction of bed porosity (packing density) based on formulations from seven authors against relative diameter.

Amin *et al.* [22] extended the effectiveness–NTU method to packed bed LTES systems, modifying the thermal resistance model for spherical encapsulation as

$$R_T = R_{\text{PCM}} + R_{\text{shell}} + R_{\text{HTF}} \tag{3.12}$$

$$R_{\text{PCM}} = \frac{1 - \lambda^{\frac{1}{3}}}{4\pi k_{\text{PCM}} \lambda^{\frac{1}{3}} r_i} \tag{3.13}$$

where r_i (m) refers to the internal diameter of the PCM sphere encapsulation and the subscript *shell* refers to the sphere encapsulation material. Amin *et al.* [21] also extended the effectiveness–NTU model to account for varying isothermal and parallel heat flows. As depicted in Figure 3.11, several parallel heat flows describe the concurrent phase change and heat transfer that occur uniformly within all spheres in the packed bed. An isothermal heat flow is used to describe the uniformity that is confined to a single horizontal layer of spheres in the packed bed. A *p-factor* is defined to determine the relative impact of each heat flow occurring in the packed bed at any instance,

$$R_T = p R_{\text{parallel}} + (1 - p) R_{\text{isothermal}} \tag{3.14}$$

$$R_{\text{parallel}} = \frac{1}{\left(\frac{4\pi r_i r_f}{r_i - r_f}\right) k_{\text{PCM}}} + \frac{1}{\left(\frac{4\pi r_i r_o}{r_o - r_i}\right) k_{\text{shell}}}$$

$$+ \frac{1}{4\pi N_{\text{spheres}} h_{\text{HTF}} r_o^2} \tag{3.15}$$

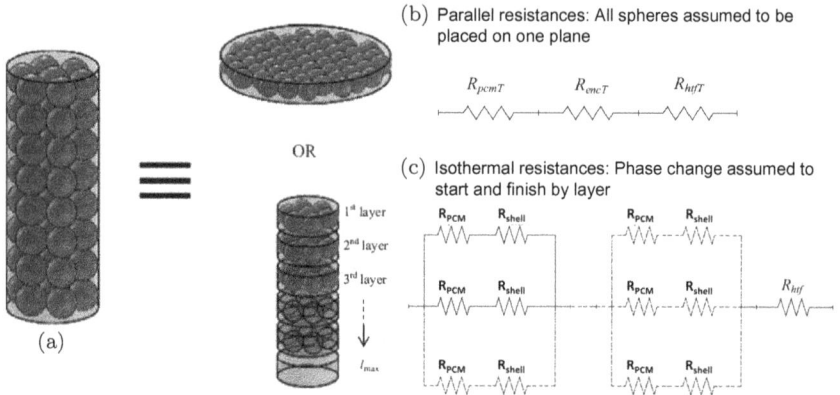

Figure 3.11.   Defining the parallel and isothermal heat flows in the packed bed [22].

$$R_{\text{isothermal}} = \frac{n_l \left(1 - \lambda^{\frac{1}{3}}\right)}{N_{\text{layer}} k_{\text{PCM}} 4\pi r_i \lambda^{\frac{1}{3}}} + \frac{n_l (r_o - r_i)}{N_{\text{layer}} k_{\text{shell}} 4\pi r_i r_o} \qquad (3.16)$$

where r_f (m) refers to the radius of the instantaneous phase-change boundary in the PCM, N_{spheres} and N_{layers} refer to the number of spheres in the packed bed and average number of spheres in a single layer, respectively. n_l refers to the total number of spheres layers in the packed bed. A p-factor value of 0.94 was found to provide the closest agreement with experimental data for both charging and discharging tests [22].

As far as numerical methods are concerned, as elaborated in Chapter 2, key classifications segment models into single or double phases [23,24]. Making appropriate modifications to the effective specific heat capacity of the encapsulated PCM spheres with formulations as described in Equations (3.1) and (3.2), the energy equation of the single-phase model can be described as

$$\varepsilon c_{p,\text{HTF}} \rho_{\text{HTF}} \frac{\delta T}{\delta t} + (1 - \varepsilon) \rho_{\text{PCM}} \frac{\delta H}{\delta t} + c_{p,\text{HTF}} \rho_{\text{HTF}} u \frac{\delta T}{\delta y}$$

$$= k_{\text{eff},y} \frac{\delta^2 T}{\delta y^2} + k_{\text{eff},r} \left(\frac{\delta^2 T}{\delta r^2} + \frac{\delta T}{2 \cdot \delta r} \right) \qquad (3.17)$$

where ε refers to the bed porosity, which can be deduced via the formulations in Table 3.1. H (kJ/m^3), u (m/s), and ρ (kg/m^3) refer to the volumetric enthalpy, axial HTF flow velocity, and density, respectively. The subscripts y and r refer to the respective axial and radial directions within the packed bed. This model is not a popular option to simulate the operations of PCM-packed beds due to the underlying assumption that the PCM and HTF are at uniform temperatures, which can only be valid for particles with very high thermal conductivity.

De Gracia and Cabeza [23] describe the three most prominent double-phase models:

i. *Concentric dispersion model*

Thermal conduction inside the PCM spheres is considered; both energy equations for solid and fluid phases are solved simultaneously; thermal gradient is assumed inside PCM, but no inter-particle heat transfer is considered.

$$\varepsilon c_{p,\text{HTF}} \rho_{\text{HTF}} \left(\frac{\delta T_{\text{HTF}}}{\delta t} + u \frac{\delta T_{\text{HTF}}}{\delta y} \right)$$

$$= \varepsilon k_{\text{HTF}} \frac{\delta^2 T}{\delta y^2} + h_{\text{convection}} \left(T_{\text{PCM}} - T_{\text{HTF}} \right)$$

$$- U_L \left(T_{\text{HTF}} - T_\infty \right) \tag{3.18}$$

$$(1 - \varepsilon) \rho_{\text{PCM}} \frac{\delta H}{\delta t}$$

$$= (1 - \varepsilon) k_{\text{HTF}} \frac{\delta^2 T}{\delta y^2} + h_{\text{convection}} \left(T_{\text{HTF}} - T_{\text{PCM}} \right) \tag{3.19}$$

$$\rho_{\text{PCM}} \frac{\delta H}{\delta t} = \frac{1}{r^2} \frac{\delta}{\delta r} \left(k_{\text{PCM}} \cdot r^2 \frac{\delta T}{\delta r} \right) \tag{3.20}$$

where $h_{\text{convection}}$ (W/m$^2 \cdot$ K) refers to the convective heat transfer coefficient between the PCM and HTF and U_L(W/m$^2 \cdot$ K) is the loss coefficient with the ambient, which can be neglected if the storage walls are assumed adiabatic. In this model, the cylindrical packed bed is axially divided into elements, where HTF temperature is assumed to be uniform and all spheres in a layer are assumed to perform as one, as shown in Figure 3.12.

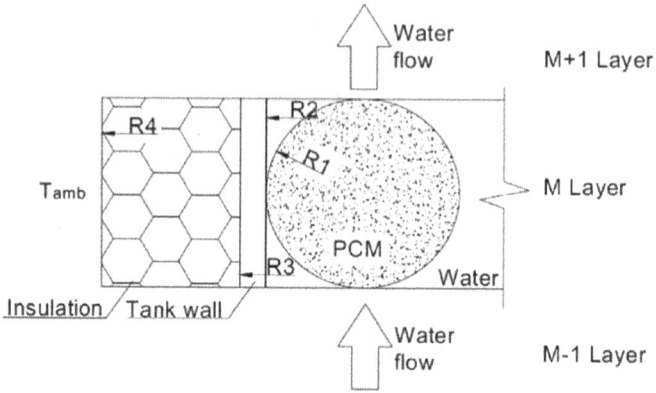

Figure 3.12.   Representation of the concentric dispersion model by de Gracia and Cabeza [23].

ii. *Continuous solid phase model*

The packed bed is considered a continuous porous medium without independent solid particles. Heat conduction in both axial and radial directions is considered for both solid and liquid phases as described in Equations (3.21) and (3.22). Hence, the HTF phase and PCM phase equations are formulated separately as

$$\varepsilon c_{p,\mathrm{HTF}} \rho_{\mathrm{HTF}} \left(\frac{\delta T_{\mathrm{HTF}}}{\delta t} + u \frac{\delta T_{\mathrm{HTF}}}{\delta y} \right) = k_{\mathrm{HTF},y} \frac{\delta^2 T_{\mathrm{HTF}}}{\delta y^2}$$

$$+ k_{\mathrm{HTF},r} \left(\frac{\delta^2 T_{\mathrm{HTF}}}{\delta r^2} + \frac{1}{r} \frac{\delta T_{\mathrm{HTF}}}{\delta r} \right) h_{\mathrm{convection}}$$

$$\times (T_{\mathrm{PCM}} - T_{\mathrm{HTF}}) - U_L (T_{\mathrm{HTF}} - T_{\infty}) \qquad (3.21)$$

$$(1 - \varepsilon) \rho_{\mathrm{PCM}} \frac{\delta H}{\delta t} = k_{\mathrm{PCM},y} \frac{\delta^2 T_{\mathrm{PCM}}}{\delta y^2} + k_{\mathrm{PCM},r}$$

$$\times \left(\frac{\delta^2 T_{\mathrm{PCM}}}{\delta y^2} + \frac{1}{r} \frac{\delta T_{\mathrm{PCM}}}{\delta r} \right) + h_{\mathrm{convection}} (T_{\mathrm{HTF}} - T_{\mathrm{PCM}})$$

$$(3.22)$$

This method permits radial variations within the bed porosity and thus the radial distribution of the HTF flow through the bed is studied using the extended Brinkman equation as conducted by Oro *et al.* [25].

iii. Schumann's model

Heat conduction is not considered for axial or radial directions but for convection between solid and liquid phases. The main limitation of Schumann's model is the inability to account for thermal dissipation inside the solid particles as well as any form of conduction.

$$\varepsilon c_{p,\text{HTF}} \rho_{\text{HTF}} \left(\frac{\delta T_{\text{HTF}}}{\delta t} + u \frac{\delta T_{\text{HTF}}}{\delta y} \right)$$

$$= h_{\text{convection}} \left(T_{\text{PCM}} - T_{\text{HTF}} \right) - U_L \left(T_{\text{HTF}} - T_\infty \right) \quad (3.23)$$

$$(1 - \varepsilon) \rho_{\text{PCM}} \frac{\delta H}{\delta t} = h_{\text{convection}} \left(T_{\text{HTF}} - T_{\text{PCM}} \right) \quad (3.24)$$

This model, however, can be adapted to account for these effects by incorporating the encapsulation and PCM thermal resistances into an overall effective thermal resistance which is written as

$$U_{\text{overall}} = \frac{1}{A} \frac{1}{R_{\text{convection}} + R_{\text{shell}} + R_{\text{PCM}} \left(\lambda \right)} \quad (3.25)$$

where A (m^2) and $R_{\text{convection}}$ (m^{2}K/W) refer to the heat transfer surface area and the convection thermal resistance between the HTF and PCM, respectively, and the internal PCM resistance, $R_{\text{PCM}}(\lambda)$ (m^{2}K/W), is expressed as a function of the solidified fraction of PCM within the capsule.

The respective convection heat transfer coefficient terms can be derived from the formulations that are listed in Table 2.3.

3.2.2.2. *Shell-and-tube LTES*

Zauner *et al.* [26] observed that, in contrast to LTES systems involving tube-side HTF flow with shell-side PCM, configurations featuring the opposite arrangement are almost completely absent in the literature. Notable exceptions to this can be found in the works of Tao *et al.* [27], Xiao and Zhang [28,29], and Zauner *et al.* [26]. Tao *et al.* [27] performed a comparative study between the design configurations of a single-tube-encapsulated PCM element and concluded that the tube-encapsulated PCM design outperformed shell-side PCM/tube-side HTF design. In addition, Tao *et al.* [27] observed

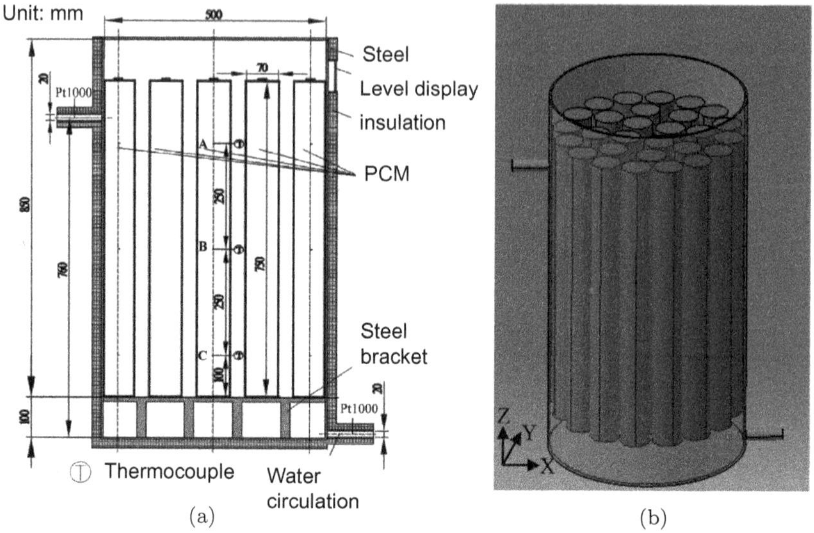

Figure 3.13. (a) Schematic representation of tube-encased PCM LTES and (b) computational model and representation by Xiao and Zhang [28]. Without flow-directing devices such as baffles within the structure, the HTF flow configuration was mostly longitudinal (parallel to the tubes) in this setup.

that a significant improvement in heat storage rate of up to 54.2% was attainable under similar operating conditions. Xiao and Zhang [28,29] conducted extensive experimental and numerical studies on the charging and discharging characteristics of an LTES containing 27 PCM tubes, as portrayed in Figure 3.13. The PCM used was paraffin incorporating varying levels of expandable graphite for heat transfer enhancement. Flow conditions were observed to strongly influence charging and discharging rates compared to inlet temperature conditions. The expandable graphite–paraffin composite PCM showed an improved heat transfer effect over pure paraffin.

Zauner *et al.* [26] incorporated the tube-encapsulated PCM into a standard shell-and-tube heat exchanger (STHE) design, complete with baffles in a forced cross-flow configuration as shown in Figure 3.14. The system employed high-density polyethylene (HDPE) as a PCM and thermal oil as an HTF for high-temperature storage operating between 110°C and 140°C. Extensive experimental work was performed along with numerical simulation and thermoeconomic optimization.

Figure 3.14.   Shell-and-tube exchanger configuration with HTF on the shell side and PCM encased within the tubes, by Zauner *et al.* [26].

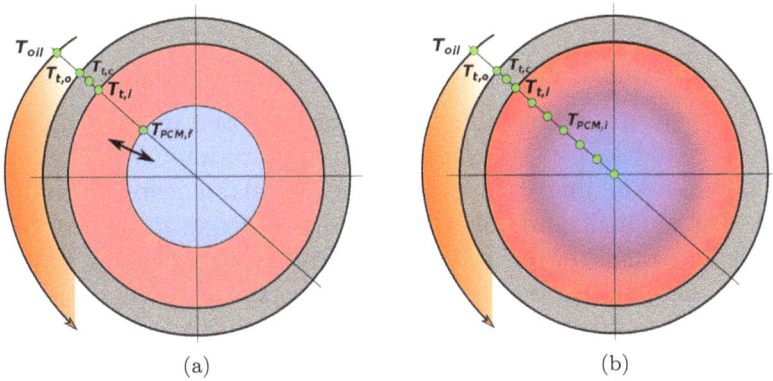

(a)   (b)

Figure 3.15.   (a) Stefan model and (b) lumped-capacitance model as tested by Zauner *et al.* for the STHE LTES simulation [26].

Xiao and Zhang [28,29] constructed a 3D computational fluid dynamics (CFD) model based on the enthalpy method. The phase-change front within the PCM was not explicitly treated and was solved based on enthalpy distribution. Zauner *et al.* [26] tested both a Stefan model and a lumped-capacitance model for the STHE LTES system as displayed in Figure 3.15. The Stefan model solves for the moving phase-change boundary in the PCM, while the lumped capacitance model incorporates the latent heat as a function of the effective specific heat capacity without solving for the phase-change boundary.

The phase-change equation for the Stefan model is formulated as

$$Q = 2\pi l h_{\mathrm{PCM}} \rho_{\mathrm{PCM}} \left(r_i - r_f \right) \frac{dr_{\mathrm{PCM}}}{dt} \qquad (3.26)$$

where l (m) refers to the length of the tube and Q(W) refers to the heat transfer rate.

The phase-change process was activated when the PCM reached the phase-change temperature. Otherwise, the PCM was modelled via a single lumped capacitance located at an intermediate position. It was deduced that $r_{\mathrm{PCM}} = 0.83 r_i$ to give the best fit with experimental data. With greater interest in this configuration of PCM encapsulation and LTES design, greater understanding of the most optimized ways to design to achieve maximum performance will prove to be valuable in the coming years.

3.2.2.3. *Slab-type encapsulation*

Slab-type encapsulations are more often deployed in building envelope TES and less in stratified TES configurations. When embedded within a building thermal envelope, the slab-encapsulated PCM allows for ambient and solar heat during the day to be stored for night use, when temperatures are lower. Conversely, in climates where cooling is imperative, cooler temperatures during the night can be used to store cold energy for warmer daytime temperatures. As HTFs are normally not required, the heat transfer between the PCM and the thermal supply/load primarily takes place through natural convection with the ambient. Shi *et al.* [30] conducted a study on slab-encased paraffin applied to a concrete wall enclosure to access its performance and impacts on indoor conditions as depicted in Figure 3.16. Key results revealed that the configuration with the PCM laminated in the wall produced the best temperature control and could reduce max temperature by up to 4°C, while the configuration featuring the PCM placed on the inner side of the wall facilitated the best humidity control.

Commercially, slab-type or panel-type encapsulations are also often utilized in consumer products such as cold carrier bags for food produce or in highly temperature-sensitive medical supplies.

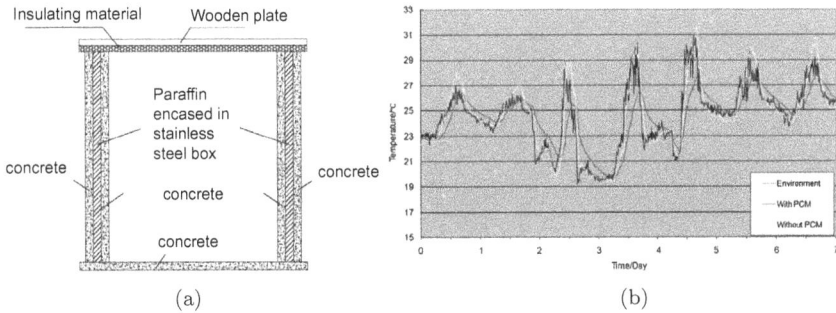

Figure 3.16. (a) Steel box-encased paraffin laminated within concrete wall and (b) the comparison between corresponding temperature profiles of the indoor, outdoor conditions and the control reference case by Shi *et al.* [30].

Figure 3.17. The Cool Cube^{TM} 28 for Lab Freezer temperatures between –50 and –15°C and the PCM panels deployed on the interior of the product to ensure internal conditions are maintained [31].

Such PCM panels can be seen deployed in low-temperature storage boxes that go to as low as –50°C, as shown in Figure 3.17 [31].

Though considerably less studied in the literature, slab-type encapsulations have also been known to be employed for industrial/commercial TES storage. PCM Products Ltd develops and deploys significant PCM slabs/panels for numerous industrial settings as shown in Figure 3.18. These tend to be more prevalent in TES systems where the PCM slabs can be stacked horizontally and HTF flows horizontally, resulting in an un-stratified TES storage. In an earlier study, Belusko *et al.* [32] introduced the effectiveness–NTU

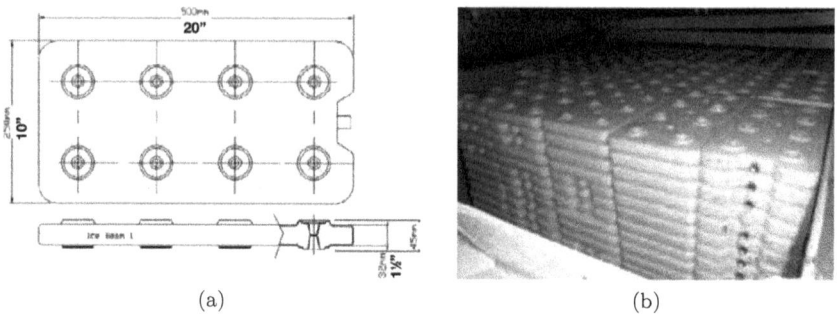

(a)   (b)

Figure 3.18.   Design (a) and in-tank deployment (b) of the FlatICE^{TM} by PCM Products Ltd. [33].

method (refer to Section 3.2.1) for layered slab-type PCM encapsulation in TES systems.

3.3.   Recent Advances in LTES Study

As highlighted in Figure 3.19, LTES studies and PCM-related publications have been attracting burgeoning interest in recent years based on a recent survey conducted by Cunha and Aguiar [34].

The following section reviews some of the most recent advances and innovations in the field of LTES and PCM study, as well as interesting commercial LTES deployments.

3.3.1.   *Academic study*

One of the key challenges with LTES deployment involves the limited number of PCMs available for each operating temperature range as well as the high cost of PCMs in general. Urban cooling temperatures for air-conditioning and mechanical ventilation (ACMV) applications span 5–12°C [35] to ensure adequate thermal comfort for indoor occupants. Due to cost, ice-based LTES systems are more widely deployed in the industries, but pose significant technical limitations including volume expansion during the freezing process and requiring sub-zero glycol chillers to charge the LTES, as highlighted in Figure 3.6. Accordingly, significant operating power and cost are incurred. Shao *et al.* [36] proposed a tailorable paraffin mixture comprising industrial *n*-tetradecane and *n*-hexadecane, tested in varying proportions

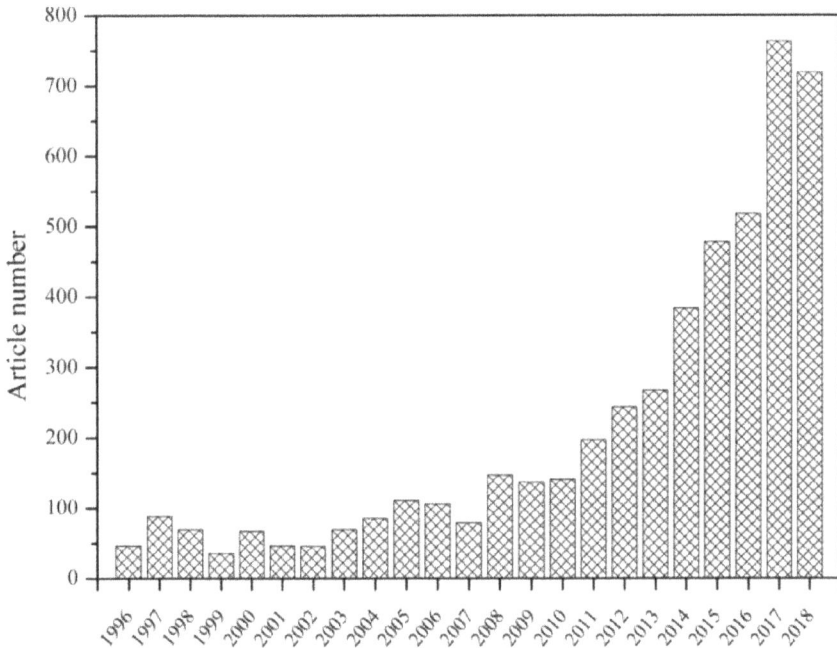

Figure 3.19.   Increasing number of PCM-related studies in recent years [34].

via the Differential Scanning Calorimetry (DSC) method. Based on the selection of mixtures tested, the mixture containing 50% n-tetradecane and 50% n-hexadecane was found to be best suited due to it possessing the lowest degree of supercooling while conceiving a phase-change temperature of 7.86°C, which lies between 5°C and 12°C (Figure 3.20).

The case study involved a commercial encapsulated ice-based LTES of height 13 m and diameter 16 m filled with spheres of 124 mm in diameter containing water. The HTF was an aqueous glycol solution (commonly termed as "brine") that retained a liquid phase as low as −4°C. To benchmark the PCM performance against the ice-based LTES system, a CFD model was developed. The model featured a sphere-column with the spheres representing the encapsulated PCM or ice and a square or hexagonal envelope representing the HTF domain as shown in Figure 3.21(a). Using operating data from the ice-based LTES system, the model was validated as illustrated in Figures 3.21(b) and 3.21(c). Comparisons made between the PCM-based and ice-based systems at 1,200 m^3/h and

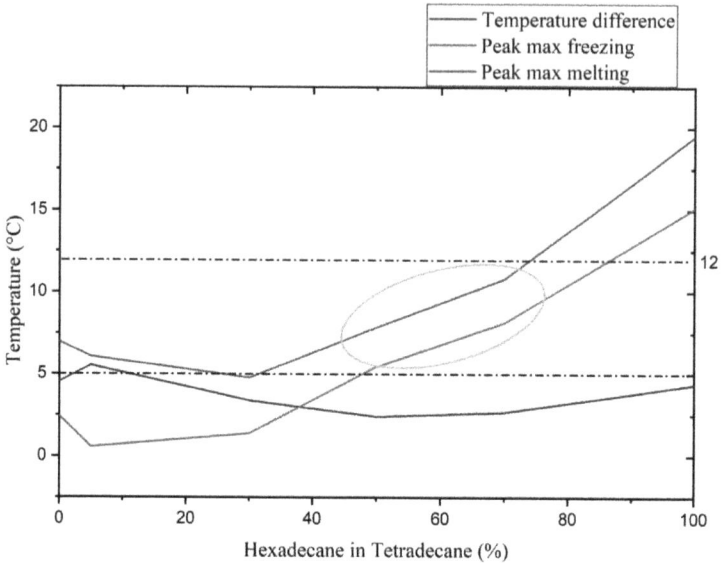

Figure 3.20. DSC-derived phase-change properties of the paraffin-based PCM mixture comprising varying proportions (wt%) of *n*-hexadecane and *n*-tetradecane as conducted by Shao *et al.* [36].

$1,400\,\mathrm{m}^3/\mathrm{h}$ HTF flow rates over 7 h demonstrated the PCM-based LTES achieving almost twice the solidification rate with an energy storage improvement of up to 9.73%.

In another application of a packed-bed PCM system, Tafone *et al.* [37] designed and experimentally investigated an alcohol-based PCM in a packed-bed cryogenic LTES for LAES cold storage. The LTES was specifically designed to improve the round-trip efficiency (RTE) of LAES systems, which are able to attain only up to 60% RTE in standalone designs. High-grade cold energy during the regasification process of liquid air in discharge can be recovered via the LTES to allow energy consumption savings on successive air liquefaction processes. An experimentally validated numerical model based on a fully implicit finite difference method was employed to simulate the behaviour of the LTES. In contrast to prior work on packed-bed systems, this packed-bed LTES featured baffles which allowed an ordered sphere-packing arrangement, as shown in Figure 3.22(a), attaining a bed porosity of 0.4 for 19 PCM spheres. Compared with a STES baseline case, the LTES packed bed was found to shorten

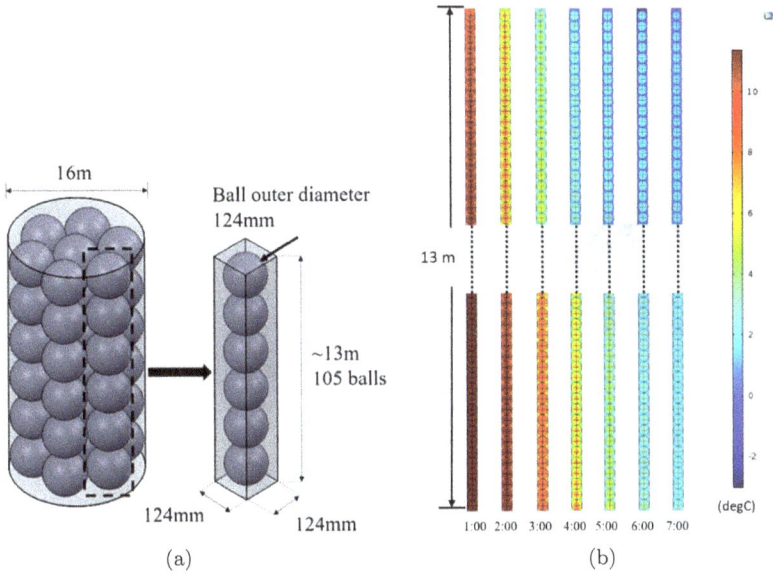

(a)

(b)

(c)

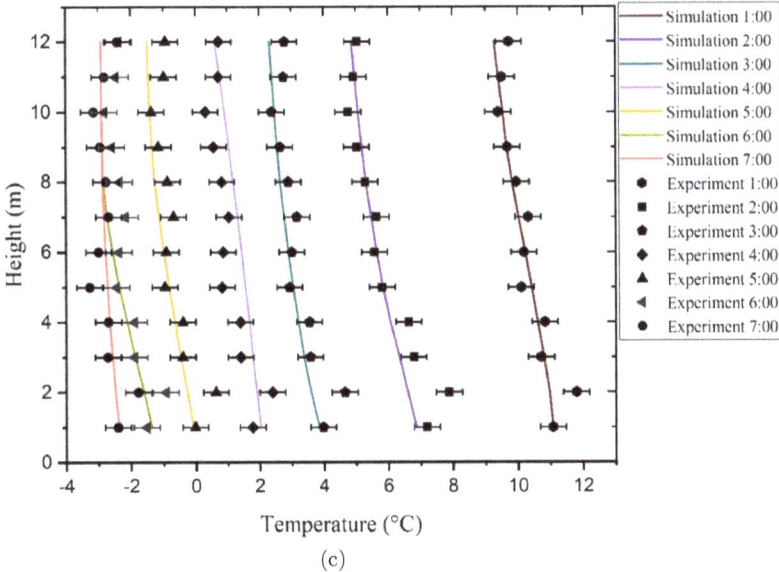

Figure 3.21. (a) Sphere-column model employed in the study. The balls represent the sphere-encapsulated ice/PCM while the rectangular envelope is adjusted to match the actual packed bed porosity of 0.47. (b) Thermocline progression of the ice-based LTES and (c) validation of the model results by comparison with operational data [36].

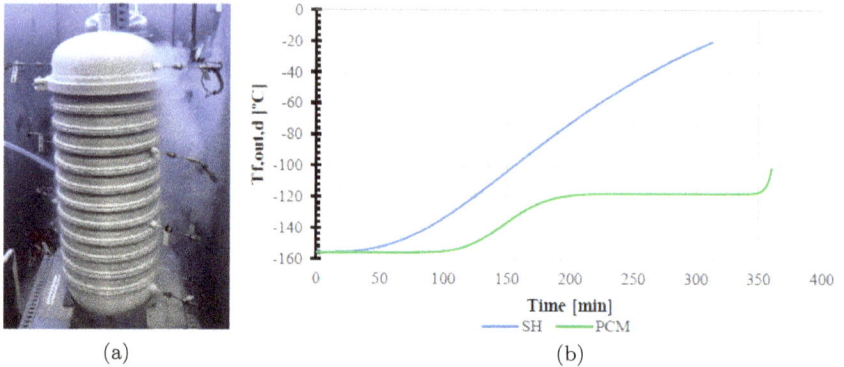

(a)   (b)

Figure 3.22.   (a) Experimental testing of the cryogenic packed-bed LTES and (b) the relative differences between the discharge temperature from the sensible TES (SH) and the Latent TES (PCM) based on computational modelling [37].

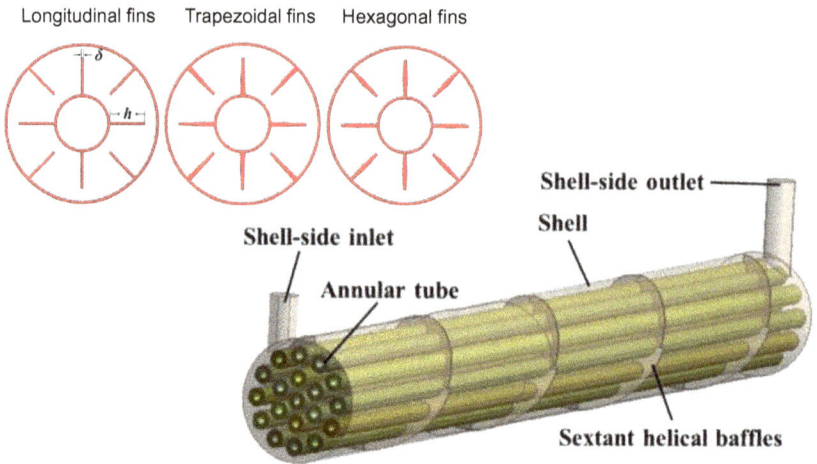

Figure 3.23.   The STHE LTES featuring sextant helical baffle and annular tubing. Various fin designs were also incorporated on the PCM side and tested [39].

charge times by 18% as well as improve the thermal stability and stratification during discharge, as is apparent in Figure 3.22(b). A payback period shorter than 3 years based on the present electrical tariffs can be achieved.

In another innovative STHE configuration, Cao *et al.* [38] and Chen *et al.* [39] proposed a sextant helical baffle design for a dual-flow tube-encapsulated LTES unit, as shown in Figure 3.23. The design is similar to the one that was studied by Zauner *et al.* [26], but features an annular tube for a second HTF flow through the centre of the PCM tubes. Additionally, a sextant helical baffle was introduced as a means to improve HTF distribution on the shell side; thereby addressing the shell-side flow non-uniformity in conventional STHE designs. Internal fins on the PCM side of the tubes were also incorporated to improve heat transfer characteristics of the LTES. A paraffin-based PCM with titanium oxide nano-additives was used for this study, while four different HTFs were tested, namely, water, fuel oil, hydraulic oil, and ethylene glycol.

Chen *et al.* [39] employed the renormalization group (RNG) k–ε model for the shell-side HTF turbulent flow, while the phase-change process was simulated using the enthalpy–porosity method. A source term, S_H, was introduced to account for the annular tube-side HTF flow, as follows:

$$\frac{\delta}{\delta t}(\rho H) + \vec{V}\nabla \cdot (\rho H) = \nabla \cdot (k\nabla T) + S_H \qquad (3.27)$$

$$H = h_0 + \Delta h_L \qquad (3.28)$$

$$\Delta h_L = \lambda h_L \qquad (3.29)$$

The numerical method was executed using FLUENT. The SIMPLE algorithm was adopted to solve the pressure and velocity coupling, while the second-order upwind scheme was used to solve the momentum and pressure equations. The tests were conducted between 25°C and 70°C with an HTF flow rate of 0.25 kg/s. Model validation was conducted with prior experimental work conducted by Cao *et al.* [38] as shown in Figure 3.24. Key findings revealed that the helix angle of shell-side HTF flow significantly impacted heat transfer. The smaller 10° helix angle was found to yield better thermal performance, while the 40° angle demonstrated an optimal outcome. For the PCM-side fin design, it was observed that the longitudinal-shaped fin of 7.5 mm height and 1 mm thickness was optimal to promote heat transfer without markedly compromising on PCM volume within the LTES. A key trade-off between the heat transfer and latent heat capacity of the LTES was also established.

Figure 3.24.   The experimental setup for the sextant helical baffle STHE LTES as tested by Cao *et al.* [38].

3.3.2.   *Industrial and commercial deployments*

Present-day TES systems in commercial thermal management systems often feature STES systems due to their simplicity in operation. Comparatively, LTES systems are less employed. Considerable efforts have been taken by the private and public sectors to bridge the gap. One district cooling plant in Singapore has installed a pilot-scale encapsulated PCM LTES system totalling a storage capacity of 120 refrigeration tonne-hours (RTh) [40]. The storage was able to supplement and store cold energy by up to 2.3 times that of a chilled water STES system of the same size, and with comparable capacities to ice-based TES systems, without needing glycol chillers to operate at sub-zero temperatures. The deployed LTES pilot unit and the encapsulated PCM balls are as portrayed in Figure 3.25.

(a)

(b)

Figure 3.25. (a) Pilot-scale LTES unit integrated with the Changi Business Park district cooling system [40] and (b) the encapsulated PCM balls used in the setup [41].

3.4.   Conclusion

In the field of thermal energy storage, studies on LTES systems are gaining significant traction and are advancing our understanding of their thermal storage capabilities. Accordingly, the industrial uptake of LTES systems is also likely to witness greater proliferation in the coming years.

In this chapter, the key fundamentals of PCM classification, behaviour, and characterization are covered. The various manifestations of LTES systems are then illustrated, particularly on the diversity of modelling methods for encapsulated LTES systems. Lastly, selected current works in the literature as well as examples related to industrial-scale deployments are described.

References

[1] G. Li, Y. Hwang, R. Radermacher and H.-H. Chun, Review of cold storage materials for subzero applications, *Energy*, vol. 51, pp. 1–17, 2013.

[2] E. Oró, A. de Gracia, A. Castell, M. Farid and L. Cabeza, Review on phase change materials (PCMs) for cold thermal energy storage applications, *Applied Energy*, vol. 99, pp. 513–533, 2012.

[3] C. Rathgeber, L. Miró, L. F. Cabeza and S. Hiebler, Measurement of enthalpy curves of phase change materials via DSC and T-History: When are both methods needed to estimate the behaviour of the bulk material in applications? *Thermochimica Acta*, vol. 596, pp. 79–88, 2014.

[4] G. Höhne, J. McNaughton, W. Hemminger and H.-J. Flammersheim, *Differential Scanning Calorimetry*, Heidelberg, Germany: Springer, 2003.

[5] J. M. Marin, B. Zalba, L. F. Cabeza and H. Mehling, Determination of enthalpy–temperature curves of phase change materials with the temperature-history method: Improvement to temperature dependent properties, *Measurement Science and Technology*, vol. 14, no. 2, pp. 184–189, 2003.

[6] SETARAM Instrumentation, SENSYS EVO TG-DSC (SIMULTANEOUS THERMAL ANALYSIS) [Online]. Available: https://www.setaram.com/setaram-products/thermal-analysis/simultaneous-thermography-differential-scanning-calorimetry-differential-thermal-analysis/sensys-evo-tg-dsc/.

[7] S. Karthikeyan and R. Velraj, Numerical investigation of packed bed storage unit filled with PCM encapsulated spherical containers — A comparison between various mathematical models, *International Journal of Thermal Sciences*, vol. 60, pp. 153–160, 2012.

[8] E. Kartashov and G. Krotov, Analytical solution of the single-phase Stefan problem, *Mathematical Models and Computer Simulations*, vol. 1, no. 2, pp. 180–188, 2009.

[9] D. Groulx, Design of latent heat energy storage systems using phase change materials, in *Advances in Thermal Energy Storage Systems*, Cambridge, UK: Elsevier, 2021, pp. 331–357.

[10] EVAPCO, Inc, Extra-Pak® Ice Coils, 2022 [Online]. Available: https://www.evapco.com/products/thermal-energy-storage/extra-pakr-ice-coils.

[11] N. Tay, M. Belusko and F. Bruno, An effectiveness-NTU technique for characterising tube-in-tank phase change thermal energy storage systems, *Applied Energy*, vol. 91, pp. 309–319, 2012.

[12] N. Tay, M. Belusko and F. Bruno, Experimental investigation of tubes in a phase change thermal energy storage system, *Applied Energy*, vol. 90, no. 1, pp. 288–297, 2012.

[13] R. Waser, S. Maranda, A. Stamatiou, M. Zaglio and J. Worlitschek, Modeling of solidification including supercooling effects in a fin-tube heat exchanger based latent heat storage, *Solar Energy*, vol. 200, pp. 10–21, 2020.

[14] T. Khadiran, M. Z. Hussein, Z. Zainal and R. Rusli, Encapsulation techniques for organic phase change materials as thermal energy storage medium: A review, *Solar Energy Materials & Solar Cells*, vol. 143, pp. 78–98, 2015.

[15] Y. Sato, T. Hirose, F. Takahashi and M. Toda, Pressure loss and liquid holdup in packed bed reactor with concurrent gas-liquid down flow, *Journal of Chemical Engineering of Japan*, vol. 6, no. 2, pp. 147–152, 1973.

[16] A. G. Dixon, Correlations for wall and particle shape effects on fixed bed bulk voidage, *The Canadian Journal of Chemical Engineering*, vol. 66, no. 5, pp. 705–708, 1988.

[17] R. Fand and R. Thinakaran, The influence of the wall on flow through pipes packed with spheres, *Journal of Fluids Engineering*, vol. 112, no. 1, pp. 84–88, 1990.

[18] E. Foumeny, H. Moallemi, C. McGreavy and J. Castro, Elucidation of mean voidage in packed beds, *The Canadian Journal of Chemical Engineering*, vol. 69, no. 4, pp. 1010–1015, 1991.

[19] R. Zou and A. Yu, The packing of spheres in a cylindrical container: The thickness effect, *Chemical Engineering Science*, vol. 50, no. 9, pp. 1504–1507, 1995.

[20] F. Benyahia and K. O'Neill, Enhanced voidage correlations for packed beds of various particle shapes and sizes, *Particulate Science and Technology*, vol. 23, no. 2, pp. 169–177, 2005.

[21] A. Ribeiro, P. Neto and C. Pinho, Mean porosity and pressure drop measurements in packed beds of monosized spheres: Side wall effects, *International Review of Chemical Engineering*, vol. 2, no. 1, pp. 40–46, 2010.

[22] N. Amin, M. Belusko and F. Bruno, An effectiveness-NTU model of a packed bed PCM thermal storage system, *Applied Energy*, vol. 134, pp. 356–362, 2014.

[23] A. de Gracia and L. F. Cabeza, Numerical simulation of a PCM packed bed system: A review, *Renewable and Sustainable Energy Reviews*, vol. 69, pp. 1055–1063, 2017.

[24] V. Palomba and A. Frazzica, Application of numerical methods for the design of thermocline thermal energy storage: Literature review and critical analysis, *Journal of Energy Storage*, vol. 46, p. 103875, 2022.

[25] E. Oró, J. Chiu, V. Martin and L. F. Cabeza, Comparative study of different numerical models of packed bed thermal energy storage systems, *Applied Thermal Engineering*, vol. 50, pp. 384–392, 2013.

[26] C. Zauner, F. Hengstberger, B. Mörzinger, R. Hofmann and H. Walter, Experimental characterization and simulation of a hybrid sensible-latent heat storage, *Applied Energy*, vol. 189, pp. 506–519, 2017.

[27] Y. Tao, Y. Liu and Y.-L. He, Effects of PCM arrangement and natural convection on charging and discharging performance of shell-and-tube LHS unit, *International Journal of Heat and Mass Transfer*, vol. 115, pp. 99–107, 2017.

[28] X. Xiao and P. Zhang, Numerical and experimental study of heat transfer characteristics of a shell-tube latent heat storage system: Part I — Charging process, *Energy*, vol. 79, pp. 337–350, 2015.

[29] X. Xiao and P. Zhang, Numerical and experimental study of heat transfer characteristics of a shell-tube latent heat storage system: Part II — Discharging process, *Energy*, vol. 80, pp. 177–189, 2015.

[30] X. Shi, S. A. Memon, W. Tang, H. Chi and F. Xing, Experimental assessment of position of macro encapsulated phase change material in concrete walls on indoor temperatures and humidity levels, *Energy and Buildings*, vol. 71, pp. 80–87, 2014.

[31] Vericor LLC, Cool Cube™ 28 at Lab Freezer Temps [Online]. Available: https://www.vericormed.com/product/cool-cube-28-varicella-medical-transport-cooler-at-frozen-temperatures-also-for-mmrv-zoster-and-ffp-ft-28/.

[32] M. Belusko, E. Halawa and F. Bruno, Characterising PCM thermal storage systems using the effectiveness-NTU approach, *International Journal of Heat and Mass Transfer*, vol. 55, pp. 3359–3365, 2012.

[33] PCM Products Ltd, Sub-zero eutectic PCM solutions, PCM Products Ltd. [Online]. Available: https://www.pcmproducts.net/files/PlusICE %20Range%202021-1.pdf.

[34] S. R. L. da Cunha and J. L. B. de Aguiar, Phase change materials and energy efficiency of buildings: A review of knowledge, *Journal of Energy Storage*, vol. 27, p. 101083, 2020.

[35] Keppel Corp, Overview, Keppel Corp, 2010 [Online]. Available: https://www.keppeldhcs.com/our_product_overview.html.

[36] Y. Shao, K. Soh, Y. Wan, M. Kumja, K. Zaw, M. Islam and K. Chua, Simulation and experimental study of thermal energy storage systems for district cooling system under commercial operating conditions, *Energy*, vol. 203, p. 117781, 2020.

[37] A. Tafone, E. Borri, L. F. Cabeza and A. Romagnoli, Innovative cryogenic Phase Change Material (PCM) based cold thermal energy storage for Liquid Air Energy Storage (LAES) — Numerical dynamic modelling and experimental study of a packed bed unit, *Applied Energy*, vol. 301, p. 117417, 2021.

[38] X. Cao, T. Du, Z. Liu, H. Zhai and Z. Duan, Experimental and numerical investigation on heat transfer and fluid flow performance of sextant helical baffle heat exchangers, *International Journal of Heat and Mass Transfer*, vol. 142, p. 118347, 2019.

[39] D. Chen, R. Zhang, X. Cao, L. Chen and X. Fan, Numerical investigation on performance improvement of latent heat exchanger with sextant helical baffles, *International Journal of Heat and Mass Transfer*, vol. 121606, p. 178, 2021.

[40] Keppel Corp, New technology to boost energy efficiency of district cooling systems, 14 October 2021 [Online]. Available: https://www. ema.gov.sg/cmsmedia/News/Media%20Release/2021/141021-Media-Release-New-Technology-to-Boost-Energy-Efficiency-of-District-Cool ing-Systems.pdf.

[41] Energy Market Authority, Smart demand side management (SMART-DSM) integration with energy efficient thermal storage system, 2021 [Online]. Available: https://energyinnovation.ema.gov.sg/files/ showcase/energy_efficiency_02.pdf.

Chapter 4

Thermochemical Thermal Energy Storage Systems

Abstract

Renewable energy sources coupled with heating and cooling applications serve as great connectors between thermal supply and demand. The previous chapters have demonstrated that sensible and latent thermal energy storage systems could be applied to *in situ* heat transfer and energy storage applications. Latent energy storage systems offer around 5–15 times higher energy storage density than sensible energy storage systems, thereby making them more compact. Principally different from sensible and latent energy storage technologies, thermochemical energy storage systems operate on reversible chemical reactions and store energy in the employed material's chemical potential. As a result, this technology offers up to 10 times more improvement in energy storage density than the latent energy storage systems. Since energy storage occurs based on the material's chemical potential, it can be stored even in ambient conditions without significant losses as long as the reactants remain separated. This characteristic offers considerable advantages in addressing the challenges associated with energy storage. This chapter discusses the fundamental operating principle of different thermochemical reactions and provides a comprehensive overview of two types of thermochemical processes: sorption and reaction-driven energy storage systems. Different designs and configurations are discussed, ideal characteristics of these materials are identified, and key challenges associated with improving this technology's readiness levels are thoroughly described.

Keywords: Thermochemical; adsorption; absorption; reaction; materials; energy storage.

4.1.   Introduction to Thermochemical TES Systems

The sensible and latent thermal energy storage systems work on the transformation of the employed material's thermophysical properties, such as the changes in the material's specific heat capacity or phase change enthalpy. In contrast, as the name suggests, thermochemical energy storage systems rely primarily on reversible chemical reactions for energy storage. Accordingly, thermochemical storage systems can be considered to be the thermal equivalent of electrochemical batteries. Unlike sensible and latent energy storage systems with energy densities ranging between 50 and 100 kWh/m^3 [1], thermochemical systems require significantly less volume as their energy density spans 200–500 kWh/m^3 [2,3]. Figure 4.1 illustrates the volume necessary to store 1,850 kWh of heat energy in different thermal energy storage materials [4]. In addition, Table 1.1 from Chapter 1 comprehensively compares thermochemical storage systems with sensible and latent counterparts. The benefits of thermochemical energy storage include reduced energy costs, high energy density, and negligible heat loss.

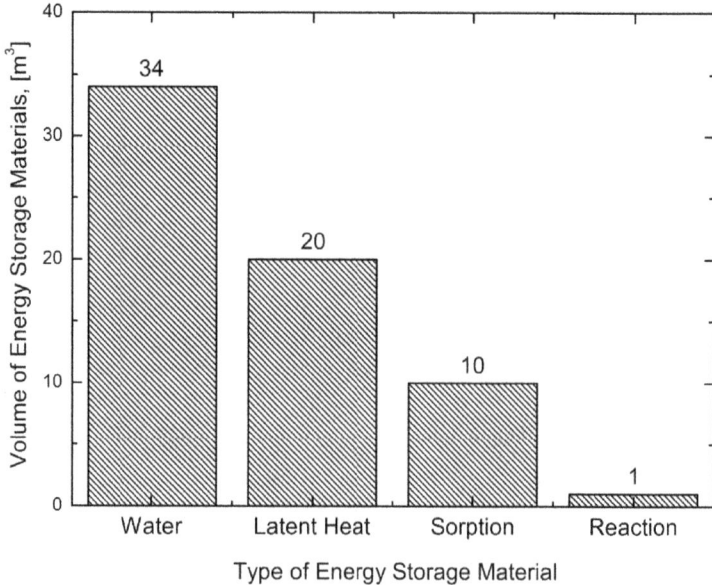

Figure 4.1.   Volume of different energy storage materials necessary to store 1,850 kWh of heat [4].

As a result, this technology can be directly applied to renewable energy applications with space constraints. Further, insignificant heat loss with the surroundings at ambient conditions implies less capital expenditure incurred for sophisticated insulation materials.

4.1.1. *Working principle*

A thermochemical energy storage system operates in three stages: charging, storage, and discharge, and is primarily governed by a reversible chemical reaction, as shown in Equation (4.1).

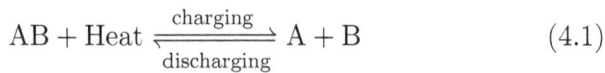

$$AB + \text{Heat} \underset{\text{discharging}}{\overset{\text{charging}}{\rightleftharpoons}} A + B \qquad (4.1)$$

In the charging step, a high-temperature heat source supplies heat to dissociate the chemical bonds of the material AB to yield products A and B. These products absorb the supplied heat and store it as their chemical potential. As long as the two products remain separated, the heat energy is stored in them without any loss, even at room temperature. Whenever energy is required, A and B are made to interact with each other to discharge the stored heat. Upon attaining the necessary reaction conditions, a reversible chemical reaction occurs to form the original product AB. The stored heat is released during the exothermic reaction. Figure 4.2 provides a schematic representation of this working principle.

4.1.2. *Classification of thermochemical energy storage systems*

The thermochemical energy storage systems are classified based on their governing mechanisms and operating temperatures. From the governing mechanism, they are categorised into sorption- and reaction-driven heat storage systems. In sorption-based systems, a reversible chemical reaction occurs due to physical or chemical absorption or adsorption (together referred to as sorption) of the sorbate molecules (usually gases) on sorbents (usually solids or liquids). Absorption is a physical or chemical process in which the absorbate molecules enter the bulk of the absorbent material. In contrast, the adsorption process occurs only at the adsorbate's surface.

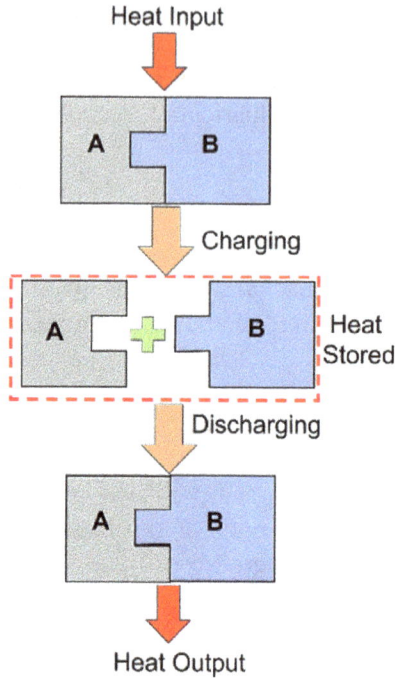

Figure 4.2.   Schematic representation of the working principle of thermochemical energy storage systems [5].

In reaction-driven systems, the reversible chemical reaction excludes sorption and involves all other chemical reactions, namely calcination–carbonation, oxidation–reduction, hydrogenation–dehydrogenation, and hydration–dehydration reactions. Due to their inherent simplicity, sorption-based systems are considered to be more developed with a higher technology readiness level (TRL) of 5–7, while the TRL of reaction-driven thermochemical storage systems is around 3–4 [6]. Aside from this, based on the operating temperature of the targeted application, the thermochemical storage systems can be classified into low temperatures ($<20°$C), low-to-medium temperatures (50–100°C), and medium-to-high temperatures (100–250°C).

Table 4.1 classifies several key thermochemical storage materials based on both operating temperature and governing mechanism and lists the types of applications they can be employed. It demonstrates the capability of thermochemical energy storage systems in both cooling and heating applications. With the significant advantages of

Table 4.1. Classification of thermochemical energy storage systems based on governing mechanisms and employed applications [7].

Operating temperature	Materials	Governing mechanism	Applications
Low temperature ($<20°$C)	Solid and liquid desiccants, salt hydrates, and metal hydrates	Sorption-based and reaction-based	Cooling, food processing, and cold storage
Low-to-medium temperatures (50–$100°$C)	Salt hydrates and metal hydrates	Reaction-based	Combined heat and power systems and space heating
Medium-to-high temperatures (100–$250°$C)	Methanol decomposition, MgO/H_2O, FeO/H_2O	Reaction-based	District heating, solar farms, and thermal storage for industrial systems

bridging the gap between energy supply and demand, thermochemical heat storage technology has the potential to become the market leader in the energy storage industry. Currently, many experimental and theoretical investigations on this technology are only at the laboratory scale. Many pilot projects and industry proof-of-concept frameworks are still in their infancy stages of development. It is worth noting that despite the advantages, thermochemical energy storage systems have not replaced sensible and latent heat storage systems.

In their current form, thermochemical storage systems have marginally reduced the impact of the energy storage problem. The limitation in their applicability emerges from their inherent drawbacks, such as high initial investments for system design, high cost of materials, poor reaction kinetics, and low cyclic stability. Appropriate resources in the form of intellectual and financial capital must be provided to make this technology feasible and economically viable. While numerous experimental and theoretical investigations focus on material synthesis, experimental analysis, and mathematical modelling, the scope of this chapter is limited to providing a comprehensive overview of the state of the art. Wherever necessary, appropriate references will be provided for readers to delve deeper into the literature and develop a more pronounced understanding of this relatively less commercialised system.

4.2.   Sorption-Driven Thermochemical Energy Storage Systems

The sorption-driven system can be solid or liquid type depending on the physical state of the material. Solid sorption materials can be further classified into three types: materials that undergo chemisorption, physisorption, and composites. Chemisorption occurs due to strong covalent bonding between sorbate and sorbent molecules and is subject to a chemical reaction. Physisorption occurs because of the weak forces of attraction: van der Waals forces or dispersion forces induced due to dipole–dipole interactions between sorbate and sorbent materials. As the physisorption process does not involve any chemical reaction, it is easily reversible with the supply of low-temperature thermal energy. In contrast, the products of the chemisorption process cannot be easily recovered with low-temperature heat sources since they lose their chemical characteristics due to the changes in their chemical bonding. This difference makes chemisorption a more energy-intensive process than the physisorption process.

In terms of applicability in thermochemical energy storage systems, the chemisorption process encompasses an extensive range of operating temperatures due to its high selectivity towards sorbate gases [8]. Further, liquid sorption is based on the concentrated solution of hygroscopic salts, such as lithium chloride (LiCl), calcium chloride ($CaCl_2$), magnesium chloride ($MgCl_2$), triethylene glycol (TEG), and ammonia (NH_3). Figure 4.3 classifies sorption-based thermochemical energy storage systems into solid and liquid systems and reviews different materials employed for each type. Further, Figure 4.4 illustrates the energy storage density of several pairs of sorption-driven materials across a wide range of operating temperatures.

4.2.1.   *Open and closed systems*

System design is one of the critical aspects of sorption-based thermochemical storage systems. Depending on whether the system

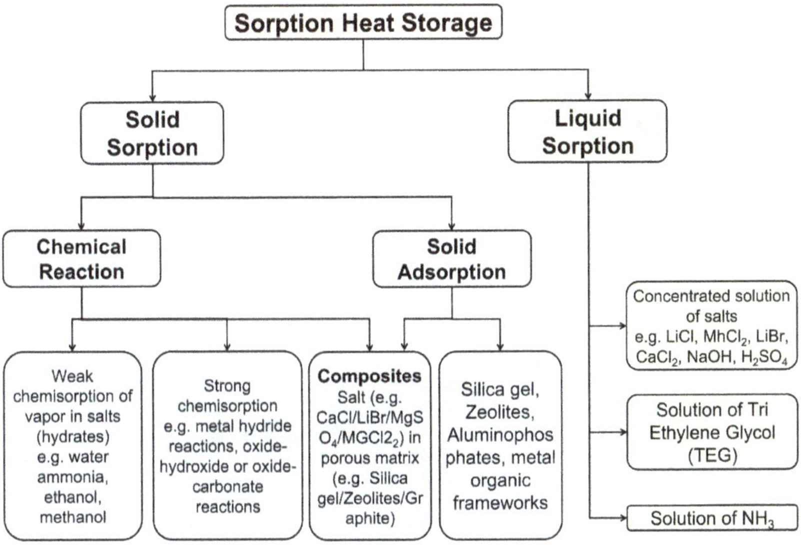

Figure 4.3. Classification of sorption-driven thermochemical energy storage systems into solid and liquid sorption [5].

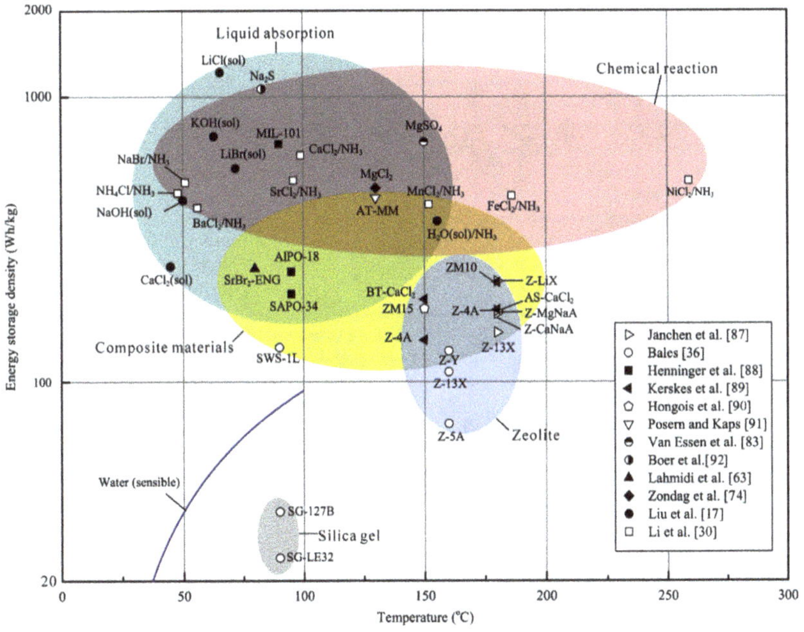

Figure 4.4. Energy storage density of different sorption-based materials at various temperatures [4].

(a)

(b)

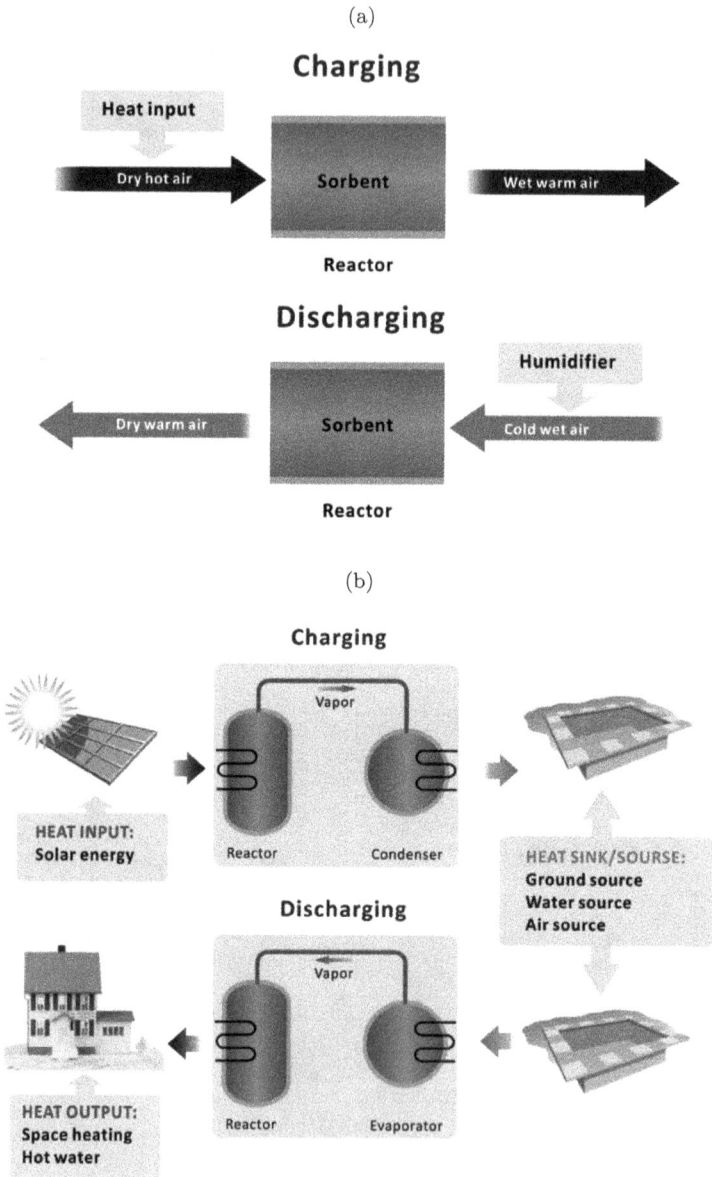

Figure 4.5. Working principle of sorption-driven thermochemical energy systems: (a) open system design and (b) closed system design [4].

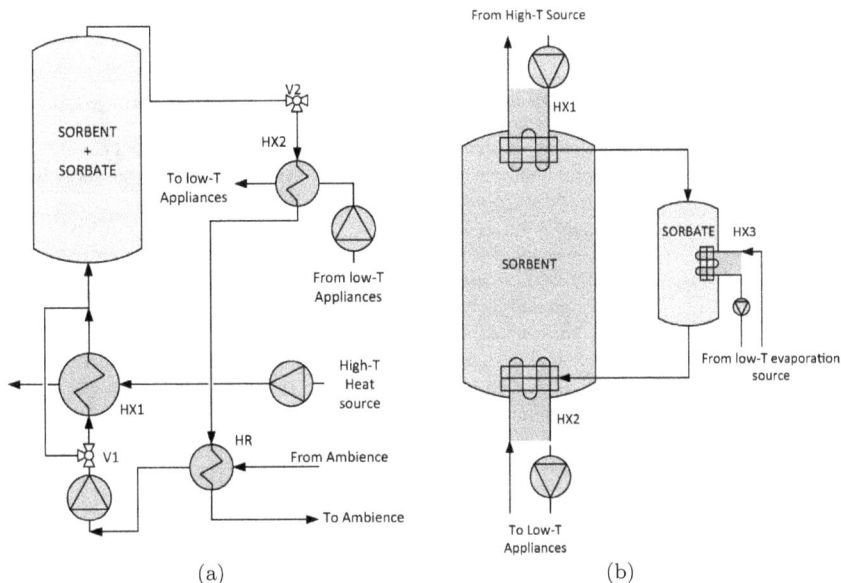

Figure 4.6. Schematic of (a) open and (b) closed system design of sorption-driven thermochemical energy storage systems [5].

exchanges mass and energy with the surroundings, sorption-driven thermochemical heat storage systems can be classified into open and closed systems. The open system operates at atmospheric pressure and exchanges both energy and mass with the surrounding environment. In contrast, a closed system design is evacuated at all times and exchanges only energy with the surrounding environment. Figures 4.5 and 4.6 illustrate the designs for open and closed systems and explain their operating principle. Additionally, Table 4.2 compares the two system designs based on their key characteristic properties and lists their advantages and limitations.

4.2.2. Absorption- and adsorption-based system configurations

This section presents the adsorbent–adsorbate pairs and some of the key system designs and configurations of sorption-based thermochemical energy storage systems. Figure 4.7 portrays a cooling system developed by integrating a vapour-compression air conditioner with an absorption-based thermochemical energy storage

Table 4.2.   A detailed comparison between open and closed system designs of sorption-driven thermochemical energy storage systems [5,7].

Characteristics	Open systems	Closed systems
Fundamental Working Principle	Operates at ambient temperature and exchanges energy and mass with the surroundings (Figure 4.5(a)).	Operates in evacuated conditions and exchanges only energy with the surroundings (Figure 4.5(b)).
Material Requirement	Must be economical, readily available, non-toxic, non-flammable, and non-hazardous.	No specific limitation on material since there is no exchanger of mass with the surroundings.
System Design Specifications	A single-bed design is sufficient; the working fluid is usually water vapour since direct exchange with the environment is involved.	Two-bed design is necessary, with the first bed consisting of sorbent and the second one acting as a condenser or evaporator to collect the liquid water.
Heat Transfer Characteristics	Higher due to simple system design.	Limited to complex heat transfer resistances.
Charging and Discharging Temperatures	Both are lower since low-temperature heat sources can be integrated.	It must be coupled with higher heat input and output sources. Therefore, it operates at higher charging and discharging temperatures.
Applications	Hot water supply and space heating.	Heat pumps, refrigeration, and energy storage.
Advantages	• Simple design and fewer physical components as compared to closed systems.	• It provides higher power density and achieves higher discharge temperatures than open systems. • Greater pressure and mass transfer control [8].
Disadvantages	• It needs a separate blower/pump to purge the sorbate flow to the surroundings. • Higher pressure drop and energy consumption for the sorbate flow. • Mass transfer is the limiting step.	• System design is costly and complicated compared to open systems. • The need for an evaporator or condenser adds up to the initial investment. • Requires frequent evacuation and maintenance to prevent undesirable accumulation of non-condensable gases. • Heat transfer is the limiting step.
Overall Efficiency	Around 70% [9].	Around 50% [9].

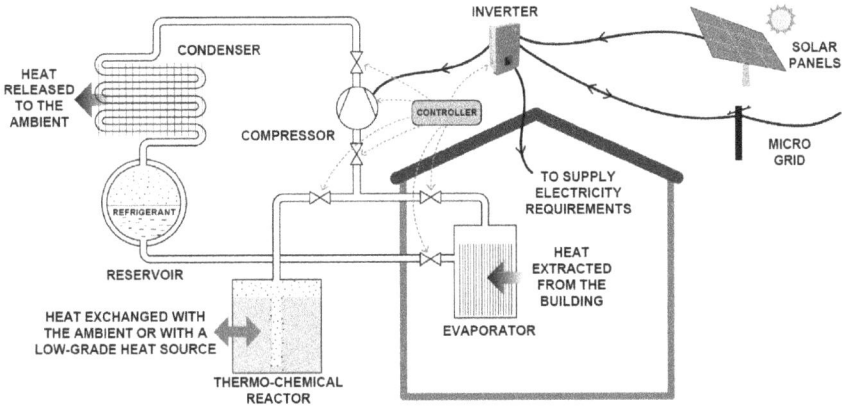

Figure 4.7. A schematic of the vapour compression refrigeration cycle integrated with a sorption-driven thermochemical energy storage system [7].

system [10,11]. The working pair is barium chloride ($BaCl_2$) salt and ammonia (NH_3) gas. The absorption-based thermochemical reactor stores energy and provides heat when no energy is available from the sun. Experimental results have demonstrated that this system is capable of producing a cooling effect of 4 kWh/day/m^2 when driven using waste heat.

The conceptual design of an open-cycle adsorption-driven thermochemical heat storage system is shown in Figure 4.8. During summer, the solar thermal air collector supplies heated air to the wet adsorbent in the reactor and regenerates it. During winter, the moist/wet air passes through the dry adsorbent bed, which releases heat to the air by absorbing its moisture. The heat released depends on the material's heat of adsorption. The selection of an appropriate adsorbent–absorbate working pair significantly depends on the required heat of adsorption in the operating range. A comprehensive review of the material characteristics of different adsorbate–adsorbent pairs has been carried out multiple times, and readers are referred to the literature [4,12].

Figure 4.9 illustrates that a similar concept of the open-cycle adsorption-based thermochemical storage system has been employed to reduce the energy demand for a district heating application in Munich [7]. This system is capable of storing up to 1,300 kWh of heat energy at the respective energy storage densities of 124 kWh/m^3 and 100 kWh/m^3 with an energy efficiency of 0.9 and 0.86 for heating and

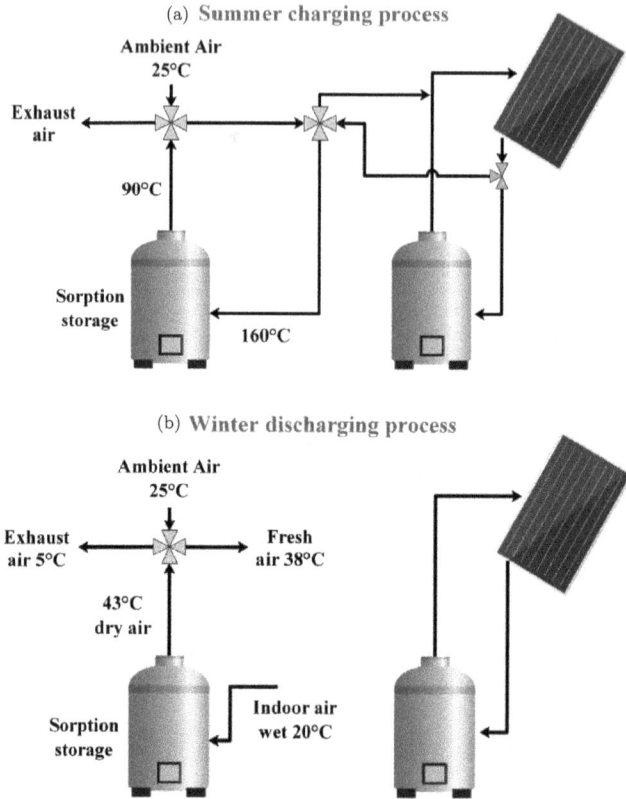

Figure 4.8.   Adsorption-based open thermochemical energy storage system with (a) energy storage step and (b) energy discharge step [7].

cooling processes. During the energy storage process at night, the zeolite adsorbent is charged, that is, thoroughly dried and regenerated at 130°C by the exhaust air from the district heating system. During the day, the heat stored in the zeolite is discharged to the building based on the required thermal energy demand. The air enters the adsorption chamber around 25–30°C and exits at around 65°C. This substantial temperature lift in the air temperature is used for heating the building.

A closed adsorption system is developed by employing silica gel/water pair for a hot water production application. Figure 4.10 illustrates that the high-temperature solar heat/waste heat is first supplied to the silica gel bed for desorption during the charging

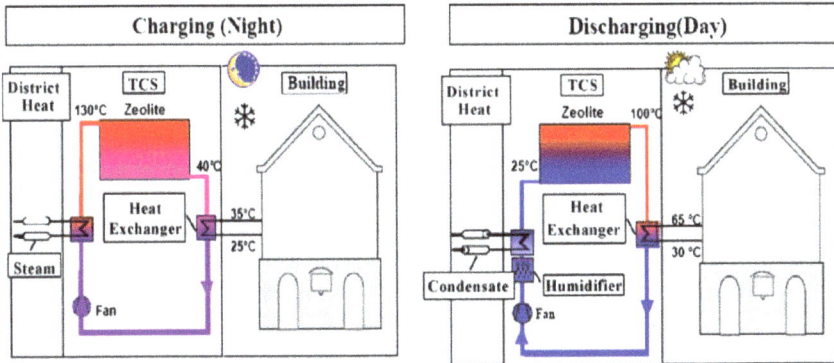

Figure 4.9. A zeolite-based adsorption-driven thermochemical energy storage application for district heating application in Munich, Germany [7].

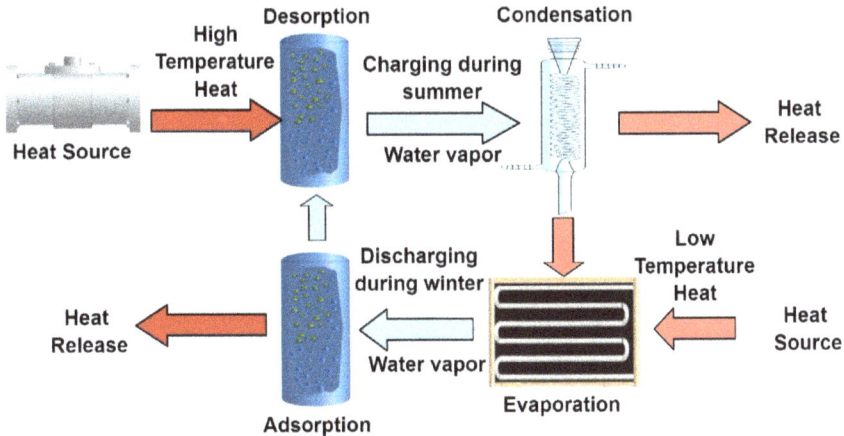

Figure 4.10. A silica gel–water-based adsorption thermochemical energy storage system for hot water production [7].

phase, and water vapour is released. Then, the water vapour is condensed and supplied to a specific application. During the discharging phase, the dry silica gel adsorbent and water vapour are made to interact with each other. The liquid water is first vaporised in an evaporator using a low-temperature heat source before passing through the silica gel. As the silica gel adsorbs the water vapour molecules, it releases the adsorption heat in an exothermic reaction.

4.3. Reaction-Driven Thermochemical Energy Storage Systems

Reaction-based systems have been applied for high-temperature thermal energy storage applications. As previously highlighted in Figure 4.1, the materials employed in this system possess significantly large energy densities and operate across wide temperature ranges. Another advantage of these reaction-driven systems over the sorption-based ones is the abundance of materials, such as salt hydrates, metal hydrides, and metal oxides. This section reviews different types of reaction-driven systems.

4.3.1. Hydration-dehydration reaction

Hydration-dehydration reactions commonly occur between metal oxides and water vapour. The commonly employed metal hydroxides for this reaction are magnesium oxide-hydroxide $(MgO/Mg(OH)_2)$ and calcium oxide-hydroxide $(CaO/Ca(OH)_2)$. These systems have high volumetric energy densities at over 900 kWh/m^3 [8], and their governing reaction equations are given by

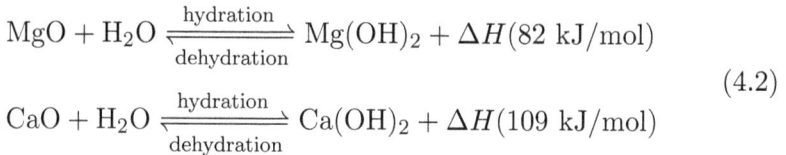

$$MgO + H_2O \underset{\text{dehydration}}{\overset{\text{hydration}}{\rightleftharpoons}} Mg(OH)_2 + \Delta H(82 \text{ kJ/mol})$$

$$CaO + H_2O \underset{\text{dehydration}}{\overset{\text{hydration}}{\rightleftharpoons}} Ca(OH)_2 + \Delta H(109 \text{ kJ/mol})$$

(4.2)

Specifically, $MgO/Mg(OH)_2$ reaction is employed as a chemical heat pump in waste heat recovery systems, as illustrated in Figure 4.11. In the heat storage mode, waste heat is supplied to dehydrate $Mg(OH)_2$, and the condensed water is stored in a tank. Whenever heat is needed, the reverse hydration reaction is made to occur at 140°C. Further, laboratory-scale tests involving this chemical heat pump design have demonstrated that the system stores surplus waste heat at around 300–400°C and can deliver it at over 200°C.

4.3.2. Redox reaction

Reduction–oxidation (redox) reactions involve the transfer of electrons. The substance that loses electrons is oxidised, whereas the

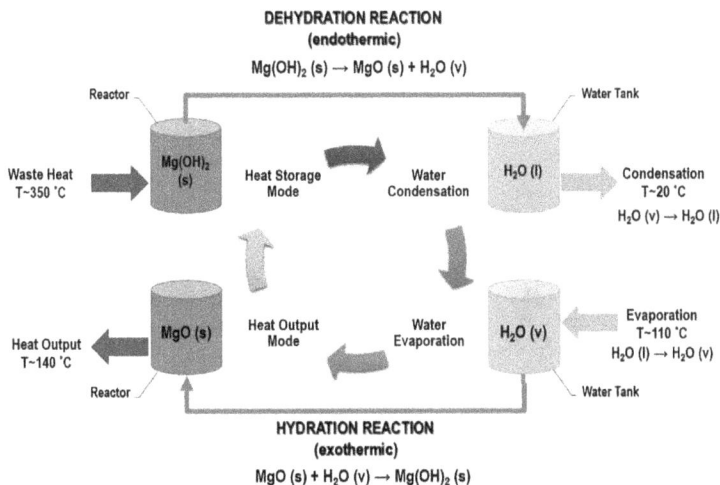

Figure 4.11. Schematic of the hydration–dehydration reaction of MgO/ Mg(OH)$_2$ occurring in a chemical heat pump [13].

compound that gains electrons is said to be reduced. Metal oxides such as barium oxide (BaO), cobalt oxide (CoO), and copper oxide (Cu$_2$O) are found to demonstrate high potential for thermochemical energy storage when they encounter oxygen (O$_2$). These reactions have been particularly explored for their thermal energy storage applications due to their extremely high volumetric energy density spanning 600–1,500 kWh/m^3. Using the oxidation reaction for thermal energy storage also removes the need to employ a gas storage system or an evaporator/condenser since the ambient air can be employed as the oxygen source as well as the heat transfer fluid. Equation (4.3) provides the oxidation–reduction reaction equations and quantifies the enthalpy released, and Figure 4.12 shows the schematic of the redox reaction occurring in Co$_3$O$_4$/CoO.

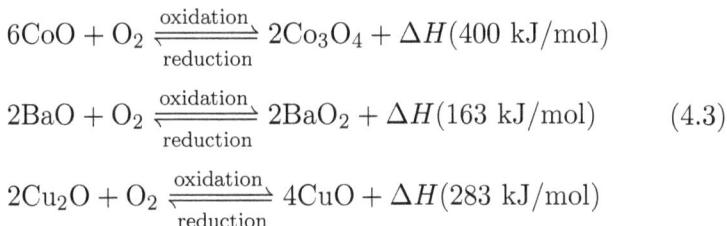

$$6CoO + O_2 \underset{\text{reduction}}{\overset{\text{oxidation}}{\rightleftharpoons}} 2Co_3O_4 + \Delta H(400 \text{ kJ/mol})$$

$$2BaO + O_2 \underset{\text{reduction}}{\overset{\text{oxidation}}{\rightleftharpoons}} 2BaO_2 + \Delta H(163 \text{ kJ/mol}) \qquad (4.3)$$

$$2Cu_2O + O_2 \underset{\text{reduction}}{\overset{\text{oxidation}}{\rightleftharpoons}} 4CuO + \Delta H(283 \text{ kJ/mol})$$

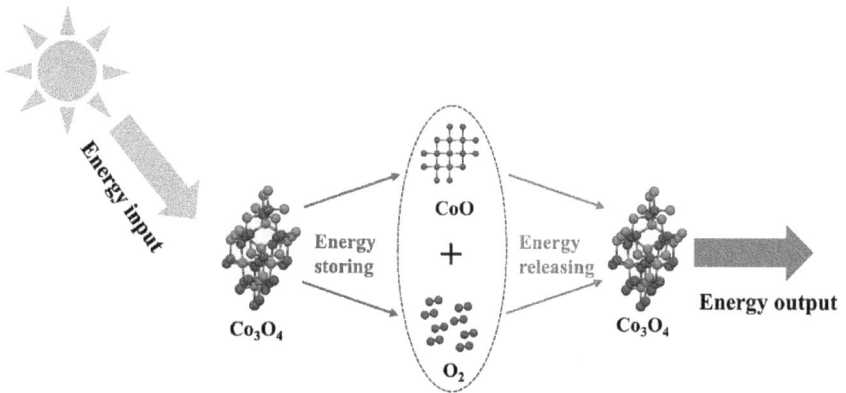

Figure 4.12.   Schematic of the Co_3O_4/CoO redox reaction [14].

Cost and toxicity are key considering parameters when selecting the materials for a redox reaction. CoO is expensive compared to other metal oxides, up to US$49,600 per ton, and it is potentially carcinogenic. Some research studies mix CoO with other chemical compounds in appropriate doping ratios to overcome its hazardous nature. As a result, both its cost and toxicity can be reduced. To realize additional benefits, mixing with materials has demonstrated excellent thermal properties that enhanced thermochemical reaction characteristics. Further, compared to CoO, the price and toxicity levels of Cu_2O and MgO are considerably lower [14].

4.3.3.   *Carbonation reaction*

Carbonation reaction involves materials reacting with CO_2 and producing carbonates, bicarbonates, and carbonic acids. This reaction type is desirable for thermal energy storage applications at temperatures greater than 800°C. The process is non-hazardous, possesses high energy density, and operates at low pressure. As represented in Equation (4.4), $CaO/CaCO_3$, $MgO/MgCO_3$, and $PbO/PbCO_3$ are some reactions studied for thermochemical energy storage applications. Their volumetric energy density typically spans 600–1,350 kWh/m^3, similar to the chemical compounds of redox reactions. It is noteworthy that significant focus has been directed explicitly towards studying $CaO/CaCO_3$ reaction for thermal energy storage because $MgO/MgCO_3$ has high refractoriness and

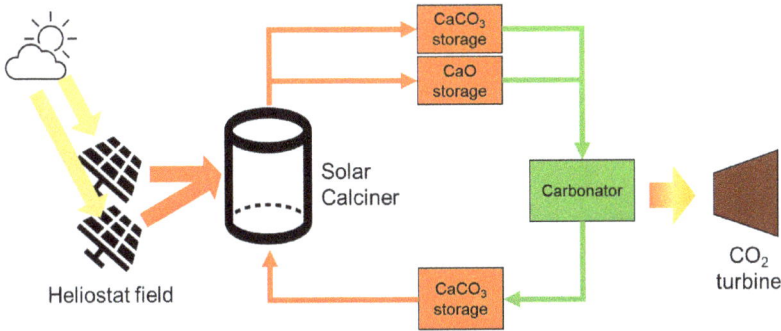

Figure 4.13. Schematic of CaO–CaCO$_3$ looping thermochemical energy storage system for concentrated solar energy plants [15].

PbO/PbCO$_3$ is toxic in nature [15].

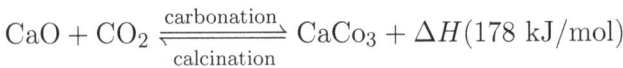

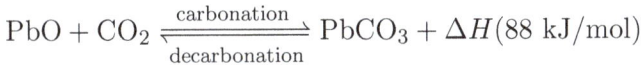

$$CaO + CO_2 \underset{\text{calcination}}{\overset{\text{carbonation}}{\rightleftharpoons}} CaCo_3 + \Delta H(178 \text{ kJ/mol})$$

$$MgO + CO_2 \underset{\text{decarbonation}}{\overset{\text{carbonation}}{\rightleftharpoons}} MgCO_3 + \Delta H(117 \text{ kJ/mol}) \qquad (4.4)$$

$$PbO + CO_2 \underset{\text{decarbonation}}{\overset{\text{carbonation}}{\rightleftharpoons}} PbCO_3 + \Delta H(88 \text{ kJ/mol})$$

Figure 4.13 illustrates a thermochemical energy storage system suitable for applications in concentrated solar power (CSP) plants based on the working principle of CaCO$_3$ calcination and CaO carbonation. In this system, the raw material CaCO$_3$ is first calcined in a calciner powered by concentrated solar energy. Calcination reaction involves heating CaCO$_3$ until it decomposes into CaO, and the carbonation reaction of CaO is an excellent way of capturing CO$_2$ from flue gases. The decomposition of CaCO$_3$ produces CaO and CO$_2$, which can be stored independently. When energy is required, CaO and CO$_2$ can be passed to the carbonator, where they generate heat, which is then used to operate a turbine and produce electricity.

4.3.4. *Ammonia decomposition reaction*

The ammonia (NH$_3$) decomposition reaction for thermochemical energy storage applications is represented by Equation (4.5). Figure 4.14 portrays the schematic of an NH$_3$ decomposition reaction

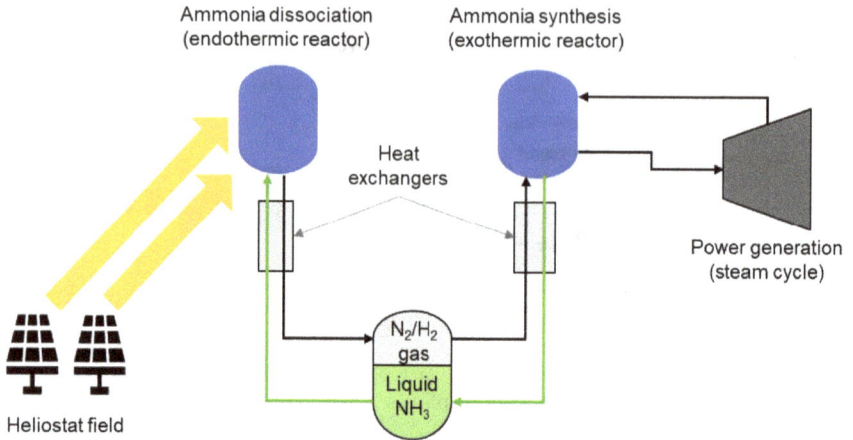

Figure 4.14. Schematic of an ammonia-based thermochemical energy storage system including ammonia dissociation, storage, and synthesis system [15].

for a solar-energy-powered thermochemical energy storage application. This system operates in a closed cycle to reduce the quantity and cost of the raw materials. NH_3 is decomposed into H_2 and N_2 gases by absorbing high-temperature solar heat during the day. High-temperature product gases enter the top chamber of the storage device and exchange heat with liquid ammonia in the bottom chamber and assist in sensible energy recovery. The entire storage system is designed as a counterflow heat exchanger to improve heat transfer rates. When N_2 and H_2 are combined to synthesise NH_3, the released heat is used to raise the temperature of the working fluid, such as supercritical steam, to drive a steam turbine and generate electricity.

$$2NH_3 + \Delta H \rightleftharpoons N_2 + 3H_2; \Delta H = 66.5 \text{ kJ/mol} \qquad (4.5)$$

4.3.5. *Methanol decomposition reaction*

Methanol (CH_3OH) is one of the cheapest and cleanest fluids available for thermochemical energy storage. Its decomposition–synthesis reaction can be employed for the storage and transportation of waste heat. When methanol is heated at 220°C and 17 bar pressure, its decomposition reaction is depicted by Equation (4.6). Figure 4.15 presents a schematic representation of a basic thermochemical energy

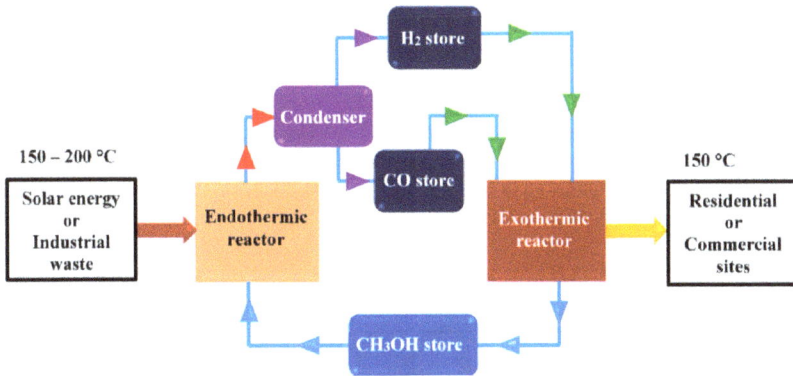

Figure 4.15. Schematic of a thermochemical energy storage system based on methanol decomposition reaction [7].

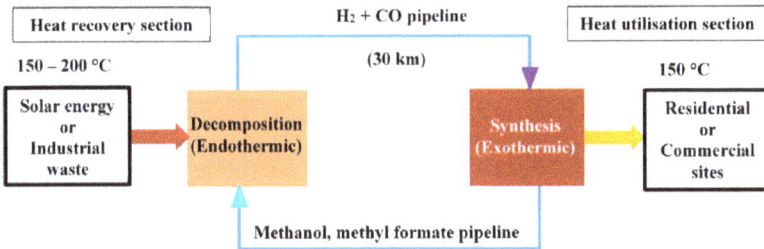

Figure 4.16. Thermal energy transport system by decomposition and synthesis of methanol [7].

storage system based on a methanol decomposition reaction. Concentrated solar power or industrial waste heat is supplied to the endothermic reactor, where CH_3OH is decomposed into syngas, a mixture of carbon monoxide (CO) and hydrogen (H_2). The produced syngas is then separated, and the individual constituents, CO and H_2, are stored separately. In the reverse combination reaction, CO and H_2 are transferred to the exothermic reactor, releasing the absorbed heat at around 150°C. In an advanced design configuration, a two-step liquid phase reaction with methyl formate (CH_3OOCH) and a one-step decomposition reaction with methanol is employed for supplying waste heat to the residential/commercial spaces located 30 km away. Figure 4.16 illustrates the schematic of this updated system, and Equation (4.7) presents the three-step chemical reaction. Mathematical modelling results for such a system demonstrate

Table 4.3. Metal hydride reactions suitable for medium-temperature thermochemical energy storage system [7].

MH chemical reaction	H_2 capacity (%)	ΔH (kJ/mol)	T_{range} (°C)
$Na_3AlH_6 + \frac{2}{3}Al + H_2 \rightleftharpoons NaAlH_4 + \Delta H$	3.73	35–38	25–202
$3NaH + Al + \frac{3}{2}H_2 \rightleftharpoons Na_3AlH_6 + \Delta H$	2.96	46–48	100–290
$2NaH + LiH + Al + \frac{3}{2}H_2$ $\rightleftharpoons LiNa_2AlH_6 + \Delta H$	3.52	55	135–315
$Na_3AlH_6 + \frac{2}{3}Al + H_2 \rightleftharpoons NaAlH_4 + \Delta H$	5.58	39	75–280

a heat recovery of over 75% [16].

$$CH_3OH + \Delta H \rightleftharpoons CO + 2H_2; \Delta H_{298K} = 90.1 \text{ kJ/mol} \quad (4.6)$$

$$CH_3OH + CO \rightleftharpoons CH_3OOCH + 38.1 \text{ kJ/mol}$$

$$CH_3OOCH + 2H_2 \rightleftharpoons 2CH_3OH + 62.8 \text{ kJ/mol} \quad (4.7)$$

$$CO + 2H_2 \rightleftharpoons CH_3OH + 95.04 \text{ kJ/mol}$$

4.3.6. *Metal hydride reaction*

Most metals and their alloys are capable of separating hydrogen (H_2) into hydrogen ions under specific temperature and pressure conditions. This process produces metal hydride (MH_n) and is accompanied by significant heat release. The general chemical reaction is given by Equation (4.8), and Table 4.3 lists complex MH reactions suitable for medium-to-high thermochemical energy storage applications.

$$M + nH_2 \rightleftharpoons MH_n + \Delta H \quad (4.8)$$

Thermochemical storage systems with MH reactions are usually closed systems with multiple MH beds. One bed is used to store energy, while the other stores hydrogen. The energy storage bed operates at the energy storage/release temperature, and the hydrogen storage bed operates at ambient temperatures. Figure 4.17 shows the working principle of an MH-based thermochemical energy storage system.

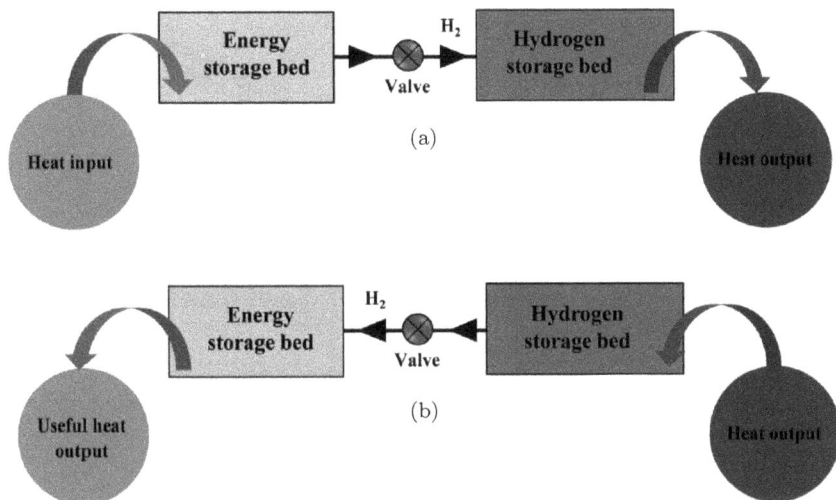

Figure 4.17. Schematic of thermochemical reaction in metal hydrides during (a) charging and (b) discharging process [7].

4.4. Ideal Material Selection Characteristics

The performance of thermochemical energy storage systems markedly depends on the characteristics of the material employed. Since the material selection affects the system's capital costs, its appropriate selection is key to engineering a thermochemical energy storage system suitable for a specific application. While numerous sorbent–sorbate pairs and salt hydrates are available, determining their suitability involves complex multi-criteria decision-making with several conflicting parameters, as illustrated in Figure 4.18. As a result, significant research efforts have been devoted to determining the most desirable thermophysical properties of the material for applications with different temperature ranges. Table 4.4 classifies the various parameters of interest into five categories: thermodynamics, kinetics, physical properties, chemical properties, and economics.

This section briefly presents a decision framework to select appropriate salt hydrates for low-temperature storage systems. The following are the kind considerations of the material employed in a thermochemical energy storage system, and Figure 4.19 lists the desired characteristics of salt hydrates for adsorption heat storage applications.

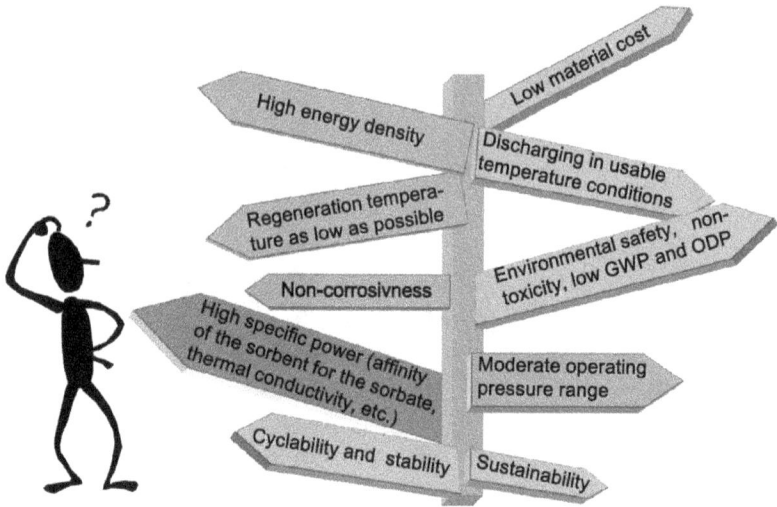

Figure 4.18.   Schematic to highlight the conflicting nature of several parameters and a need for a trade-off in material selection of thermochemical energy storage systems [17].

(1) The charging and discharging temperatures should be in the operating range of the target application.

(2) The volumetric storage density determines the storage system's size, and the storage density must always be greater than the application requirement.

(3) The number of moles of water released. The dehydration enthalpy of salts spans between 54 and 73 kJ/mol. Salts that release large moles of water are preferred.

(4) The price per kWh of energy stored. A low price per kWh is desired.

(5) The material properties should include non-degradable, non-toxic, non-flammable, non-explosive, environmentally friendly, and readily available.

The material screening method must be simple, include an exhaustive list of numerous possible materials, and provide results in a short time frame. Accordingly, a three-step process is suggested by Tsoukpoe and co-researchers [17], which involves a simple elimination method based on safety characteristics. The following steps

Table 4.4.   Summary of desired parameters required for selecting an appropriate material for thermochemical energy storage systems [5].

Categories	Parameters
Thermodynamic	Charging/Desorption temperature
	Discharging/Sorption temperature
	Heat input
	Heat output
	Storage efficiency
	Regeneration time
	Sorbate's mass transit
	Non-reactive to O_2
Physical	Density
	Operating pressure
	Degree of sorbate loading (%)
	Volume change (%)
Kinetic	Reaction kinetics (m/s)
Chemical	No chemical decomposition
	Non-corrosive
	Non-flammable
	Non-toxic
	Non-explosive
Economic	Abundant
	Inexpensive

employ material characteristic data from the combined Thermogravimetric Analysis (TGA) and Differential Scanning Calorimetry (DSC) measurements for comparison. In TGA, the material's thermal stability is evaluated by monitoring its mass change as the sample is heated at a constant rate. DSC is a thermal technique where the sample is exposed to a controlled temperature environment, and the heat flowing into or out of a sample is measured as a function of temperature or time. These thermal analysis techniques are usually employed to compute melting/boiling phases, mass changes, kinetics, and transition enthalpies.

The three steps have been described in detail, as portrayed in Figure 4.20, for a specific case of selecting the best salt hydrates to store 80 kWh of thermal energy in a storage unit of 1 m³ at a charging temperature of 105°C and discharging temperature of 60°C

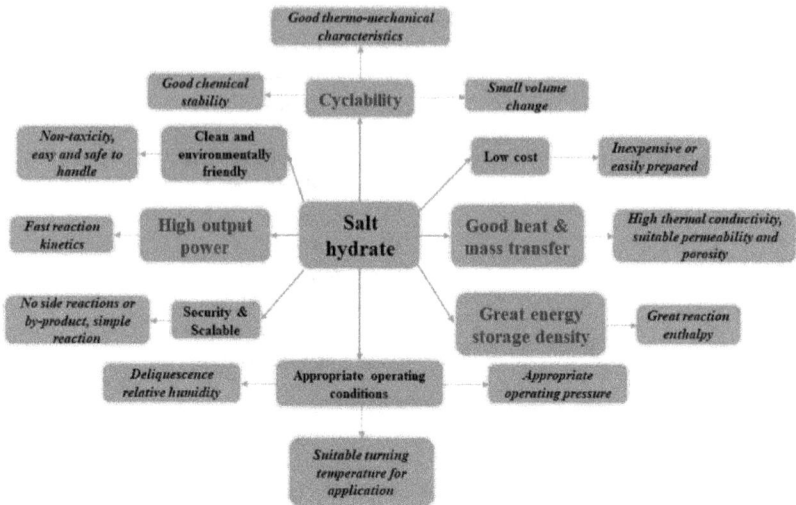

Figure 4.19.   Selection criteria for salt hydrate-based adsorption thermochemical materials [18].

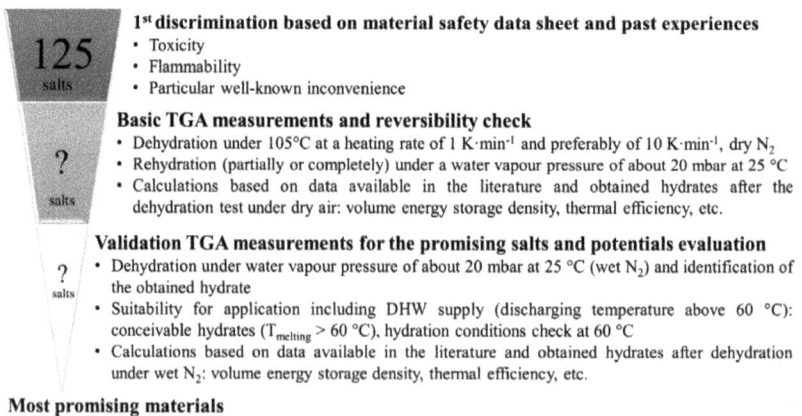

Figure 4.20.   A three-step screening procedure for selecting salt hydrates [17].

for a domestic hot water production application. 125 salt hydrates were identified in the first step, and their material safety data sheets (MSDS) were scrutinised. Any material with a health hazard rating above three according to the NFPA 704 standard was eliminated. In addition, factors such as flammability and explosiveness were also checked, and materials exhibiting these properties were rejected.

Lastly, any selected material that demonstrated oxidising behaviour was rejected for the purpose of achieving long-term cyclic stability.

In the next step, the heating rate and rehydration stability of the material were checked via TGA measurements. This step should ideally be based on the type of application, its operating condition, and the desired heating rate. Candidates with high energy storage density and thermal efficiency were selected for further screening through DSC measurements. In this last step, the material's suitability to discharge heat at a particular temperature should be accounted for.

4.5. Conclusions

Thermochemical energy storage offers an energy-efficient solution to mitigate the mismatch between energy generation of renewable systems and peak energy consumption. Additionally, it can store a large quantity of heat for both long- and short-term durations even at ambient temperature, with minimal heat loss. Further, due to the nature of storing energy in the material's chemical potential, thermochemical energy storage systems have 5–10 times higher energy density than sensible and latent energy storage systems.

Thermochemical energy storage systems can be broadly classified into two types based on the governing mechanisms: sorption- and reaction-driven systems. Solid adsorption, liquid absorption, and composite-based sorption reactions fall under the sorption-driven systems, whereas reaction systems encompass other chemical reactions, such as hydration–dehydration, decomposition–synthesis, redox, and carbonation–decomposition/calcination. This chapter has comprehensively documented the different materials employed, system designs and configurations, and the type of applications it can be employed. Further, it discusses the desired material characteristics required for being employed in a thermochemical energy storage system and introduces a three-step approach to determine the best available material. Despite many advantages, this thermal storage technology is less applied in the commercial settings because it involves high initial capital cost and poor long-term stability. As a result, the focus of this technology is to continue to conduct research and development in materials development and lab-scale prototyping levels. With advancements in material science and improvements in heat exchanger/bed designs, this technology would eventually become economically viable for wide-scale deployment.

References

[1] D. Lefebvre and F. H. Tezel, A review of energy storage technologies with a focus on adsorption thermal energy storage processes for heating applications, *Renewable and Sustainable Energy Reviews*, vol. 67, pp. 116–125, 2017. https://doi.org/10.1016/J.RSER.2016.08.019.

[2] D. Alezi, A. Shkurenko, U. Samin, H. Aggarwal, Ł. J. Weseliński, Y. Belmabkhout, *et al.*, Reticular chemistry in action: A hydrolytically stable MOF capturing twice its weight in adsorbed water, *Chem*, vol. 4, pp. 94–105, 2018. https://doi.org/10.1016/j.chempr.2017.11.005.

[3] A. Karmakar, V. Prabakaran, D. Zhao and K. Jon, A review of metal-organic frameworks (MOFs) as energy-efficient desiccants for adsorption driven heat-transformation applications, *Applied Energy*, vol. 269, p. 115070, 2020. https://doi.org/10.1016/j.apenergy.2020.115070.

[4] N. Yu, R. Z. Wang and L. W. Wang, Sorption thermal storage for solar energy, *Progress in Energy and Combustion Science*, vol. 39, pp. 489–514, 2013. https://doi.org/10.1016/j.pecs.2013.05.004.

[5] K. Kant and R. Pitchumani, Advances and opportunities in thermochemical heat storage systems for buildings applications, *Applied Energy*, vol. 321, p. 119299, 2022. https://doi.org/10.1016/j.apenergy.2022.119299.

[6] L. André and S. Abanades, Recent advances in thermochemical energy storage via solid -gas reversible reactions at high temperature, *Energies*, vol. 13, p. 5859, 2020. https://doi.org/10.3390/en13225859.

[7] F. Desai, J. Sunku Prasad, P. Muthukumar and M. M. Rahman, Thermochemical energy storage system for cooling and process heating applications: A review, *Energy Conversion and Management*, vol. 229, p. 113617, 2021. https://doi.org/10.1016/j.enconman.2020.113617.

[8] H. Bao and Z. Ma, *Thermochemical Energy Storage*, Oxford, UK: Elsevier Inc., 2022. https://doi.org/10.1016/B978-0-12-824510-1.00028-3.

[9] C. Tregambi, P. Salatino, R. Solimene and F. Montagnaro, An experimental characterization of Calcium Looping integrated with concentrated solar power, *Chemical Engineering Journal*, vol. 331, pp. 794–802, 2018. https://doi.org/10.1016/J.CEJ.2017.08.068.

[10] B. Michel, N. Mazet, S. Mauran, D. Stitou and J. Xu, Thermochemical process for seasonal storage of solar energy: Characterization and modeling of a high density reactive bed, *Energy*, vol. 47, pp. 553–563, 2012. https://doi.org/10.1016/J.ENERGY.2012.09.029.

[11] F. Ferrucci, D. Stitou, P. Ortega and F. Lucas, Mechanical compressor-driven thermochemical storage for cooling applications in tropical insular regions. Concept and efficiency analysis, *Applied Energy*, vol. 219, pp. 240–255, 2018. https://doi.org/10.1016/J.APENERGY.2018.03.049.

[12] K. Thu, N. Nasruddin, S. Mitra and B. B. Saha, Sorption-based energy storage systems: A review, *Makara Journal of Technology*, vol. 23, p. 16, 2019. https://doi.org/10.7454/mst.v23i1.3691.

[13] E. Mastronardo, L. Bonaccorsi, Y. Kato, E. Piperopoulos and C. Milone, Efficiency improvement of heat storage materials for $MgO/H_2O/Mg(OH)_2$ chemical heat pumps, *Applied Energy*, vol. 162, pp. 31–39, 2016. https://doi.org/10.1016/j.apenergy.2015.10.066.

[14] X. Han, L. Wang, H. Ling, Z. Ge, X. Lin, X. Dai, *et al.*, Critical review of thermochemical energy storage systems based on cobalt, manganese, and copper oxides, *Renewable and Sustainable Energy Reviews*, vol. 158, p. 112076, 2022. https://doi.org/10.1016/j.rser.2022.112076.

[15] D. Liu, L. Xin-Feng, L. Bo, Z. Si-quan and X. Yan, Progress in thermo-chemical energy storage for concentrated solar power: A review, *International Journal of Energy Research*, vol. 42, pp. 4546–4561, 2018. https://doi.org/10.1002/er.4183.

[16] L. J. Mauer and L. S. Taylor, Water-solids interactions: Deliquescence, *Annual Review of Food Science and Technology*, vol. 1, pp. 41–63, 2010. https://doi.org/10.1146/ANNUREV.FOOD.080708.100915.

[17] K. E. N'Tsoukpoe, T. Schmidt, H. U. Rammelberg, B. A. Watts and W. K. L. Ruck, A systematic multi-step screening of numerous salt hydrates for low temperature thermochemical energy storage, *Applied Energy*, vol. 124, pp. 1–16, 2014. https://doi.org/10.1016/j.apenergy.2014.02.053.

[18] W. Li, J. J. Klemš, Q. Wang and M. Zeng, Salt hydrate-based gas-solid thermochemical energy storage: Current progress, challenges, and perspectives, *Renewable and Sustainable Energy Reviews*, vol. 154, p. 111846, 2022. https://doi.org/10.1016/j.rser.2021.111846.

Chapter 5

Process Integration for Thermal Energy Storage Systems

Abstract

The key purpose of a TES unit is simply to perform as a time-shifting buffer between peak supply and demand periods over varying timescales. However, the integration of TES units into a thermal operating system is non-trivial and involves several issues, including (1) How should they be sized? (2) How should they be optimized to suit the needs of the end user? and (3) How should they be incorporated into a system in order to promote the overall system performance? This chapter describes system-level TES integration and the various approaches taken by the industrial and academic fields to incorporate TES as an integral part of system-level operation. Fundamental principles such as the system operating design and control will be first covered, followed by recent and more niche applications in the various industrial sectors.

Keywords: Process integration; LNG regasification; smart systems; optimization.

5.1. Process Integration of TES Systems

Gibb *et al.* [1] highlighted that the diverse complexity of process integration for TES systems is due to the wide range of situations in which TES can be applied, even within the same industry. The type of energy required or provided ranges from heating, cooling, and in some cases electricity. A key challenge involves identifying the factors that make the differing TES advantageous and/or disadvantageous and matching these TES with the specific applications. Gibb *et al.* [1] further characterized process integration of TES into two distinct types: greenfield or brownfield (retrofitted). In the former, the TES is designed and sized with the rest of the process from the beginning stage. In the latter, the TES is sized to fit the needs of an existing ongoing process. As shown in Figure 5.1, the relationship between a TES and the process it serves is defined by the mutual requirements each needs to meet. This allows the technoeconomic boundaries of the overall system to be characterized. Accordingly, the key performance indicators (KPIs) of the integrated technology can then be determined.

5.1.1.   *The TES — Process integration methodology*

Three key features of the process and the TES during the characterization phase include (1) process, (2) storage, and (3) storage integration with the process. The feasibility of TES process integration is firstly dependent on the process requirements that the TES has to first fulfil, and secondly on the performance of the TES under the process conditions. In other words, the application provides the context for which the TES integration can be assessed.

5.1.1.1.   *Process analysis*

In the process analysis step, all information relating to the integration of the TES needs to be comprehensively addressed. This allows the complex relationship of the TES with the process to be simplified into several key points. The key steps of this process, depicted in Figure 5.2, are detailed as follows:

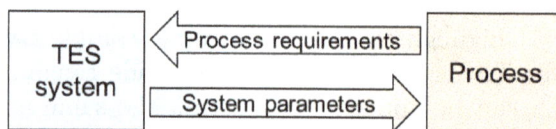

Figure 5.1.   Relationship between TES and process [1].

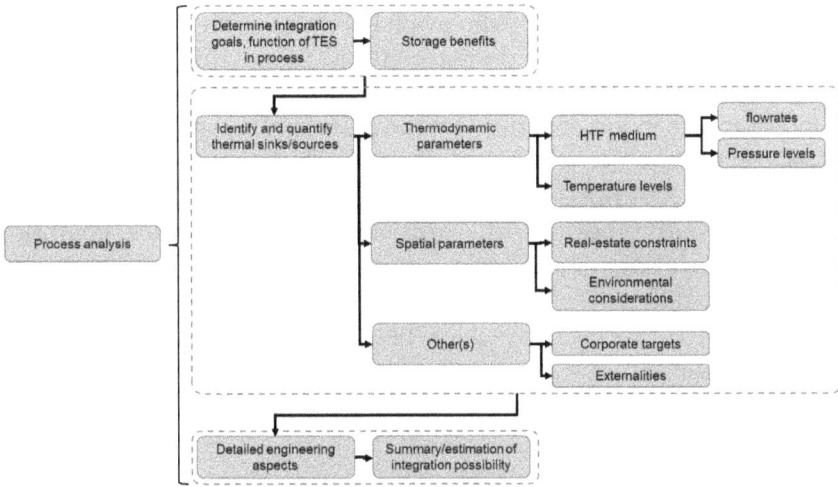

Figure 5.2. The process analysis methodology for TES integration as proposed and structured by Gibb *et al.* [1].

1. *Structured collection and analysis of process information*

   First, the goal of integration is established, while detailing the function of the TES within the system/process. It is crucial to delineate the boundaries of the process and the storage system at this stage in order to determine the factors that can and need to be analysed.

2. *Planning and evaluation of approximate storage plan*

   Next, the thermal sinks and sources within the boundaries of the process are determined. This step involves consultation with subject matter experts such as specialists as well as obtaining available information regarding the process, such as P&ID (piping and instrumentation diagram). Thermal sinks include examples such as heat pumps, process heat, liquefied natural gas (LNG) regasification, and/or power cycles. Thermal sources include urban cooling loads, waste heat, combustion, and/or even solar radiation [1]. Quantifying these sources and sinks then allows for the evaluation of the TES integration potential. Thereafter, two key parameter types will have to be evaluated. The first types are the thermodynamic parameters, which include (1) heat transfer medium/fluid (HTF), (2) temperature levels, and (3) transient profiles. The HTF from the thermal sink/source influences the thermal transfer rate, containment type, the available storage medium/materials, and the overall storage design/concept. Temperature levels of the source/impact the thermal flows and capacities of the system. These in turn are governed by the transient thermal profiles from the sources/sinks, which are also strongly influenced by the mass flow rates and pressure levels of the HTF.

The second types are the *spatial parameters*, which comprise details relating to available space and distances and obstacles between different parts of the process, as well as the installed infrastructure. Once these have been quantified, the non-technical issues can then be addressed. These include the process recurrence, corporate targets, environmental considerations, and other externalities.

3. *Detailed engineering aspects*

Once all the above have been addressed, the integration possibilities can then be estimated and analysed.

5.1.1.2.   *Defining the TES system*

Central to assessing the integration of the TES into a process is the establishment of the boundaries of the TES system relative to the process. Due to the diversity of TES applications, there is widespread disagreement on where the system boundaries should be established. To address this, two distinct aspects of TES systems must be established. These are components and modules. Components refer to the smallest parts of the TES system. Modules constitute a collection of components that serve a specific task/role within the TES system.

According to Gibb *et al.* [1], the TES system boundary is the point of contact between the fluid streams and the thermal sink and thermal source. The system contains all the components and modules exclusively used by it and those necessary to deliver heat to the sink and to retrieve heat from the source.

Accordingly, the TES system boundary can be established to only encompass the components and modules that allow it to serve its purpose of charging and discharging thermal energy. An example of this is a TES-integrated concentrating solar plant (CSP) as illustrated in the schematic representation in Figure 5.3.

In Figure 5.2, a thermal source and sink are defined by units 1 (solar field) and 2 (power block). The TES system, connected in

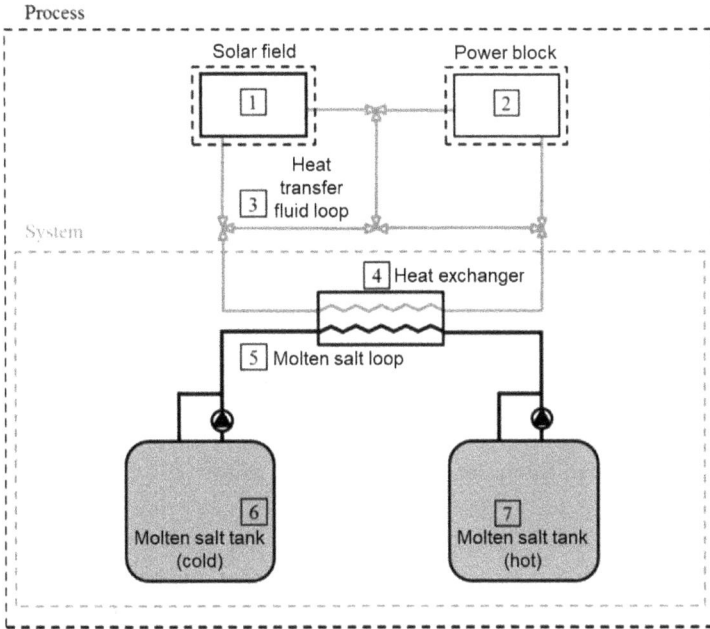

Figure 5.3. Defining the TES system boundaries for integration with a CSP [1].

parallel to the CSP, consists of units 4 (heat exchanger), 5 (molten salt loop between storage tanks and heat exchanger), as well as 6 and 7 (the cold and hot molten salt storage tanks). These 4 units are necessary to deliver and receive the thermal energy to and from the process and, hence, are defined within the TES boundary. Unit 3, which defines the heat transfer fluid loop, however, is not considered a part of the TES system as it would need to exist regardless of the storage.

5.1.1.3. *Evaluating the performance of the TES integration — Key performance indicators*

The penultimate step in the methodology involves assessing the benefits brought to the process via the TES integration. As explained earlier, the application contextualizes the performance of a TES, which on its own cannot be evaluated without a set purpose. Key performance indicators (KPIs) are defined to establish the benefits of the TES. These performance indicators refer to the specific TES

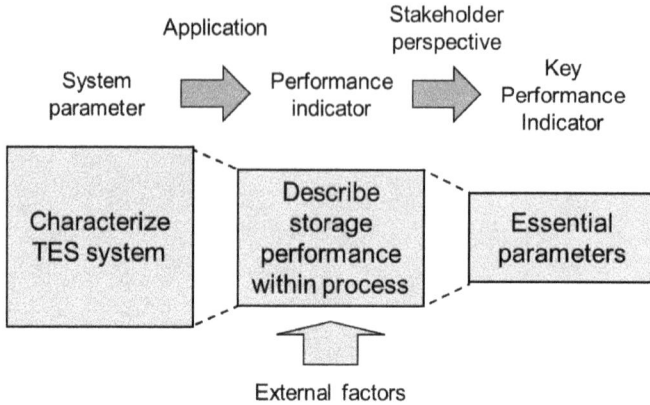

Figure 5.4. Filtering the performance parameters of the TES to the essential key performance indicators for a comprehensive assessment of the process integration [1].

system parameters that are essential to meet process requirements. This can take the form of the power delivered by the TES during discharge. Beyond the technoeconomic parameters of the TES system, qualitative external factors such as process/grid flexibility also contribute to the impact that the TES system makes on the process. To comprehensively assess the impact of the TES, differing performance indicators must be prioritized according to the diversity of stakeholder views. Stakeholders are parties with the most invested interests in the entire integration exercise. They identify and select interests that influence the outcome of the TES integration. This narrowing of the performance parameters to several key performance indicators can be summarized in Figure 5.4. Each successive step of evaluation provides a unique assessment of the TES integration while facilitating a comprehensive assessment from all angles.

5.1.2. *Economic analysis*

One key aspect of process integration that needs to be considered is the suitable economic analyses to enable accurate costing of the system. Conventionally, numerous methods to approximate costing of water or ice storage tanks are available in the literature [2,3]. Some of the costing parameters can be adopted directly from storage tank suppliers, or estimation using semi-empirical equations based on the reviews of past studies. Sebzali *et al.* [4], for instance, performed an economic analysis for chilled water storage tank systems as a part of an air-conditioning system. A steel tank costing method was adopted using input from a commercial supplier for an in-situ constructed cylindrical steel tank that was able to withstand up to 500 kPa pressure. It was further assumed that insulation and cladding constituted about 12% of the tank cost under load-levelling strategy, 50% for demand-limiting strategy, and 10.9% for full-storage strategy.

In another study that examines the feasibility of ice storage tanks in Saudi Arabia, Habeebullah [2] employed a method adopted from the commercial supplier Baltimore Aircoil Company [5], which supplies ice-on-coil latent TES (LTES) systems.

$$Z_{ST} = \alpha_{ST} E_{ST} \tag{5.1}$$

where Z refers to the cost and the subscript ST refers to the TES. E (kWh) refers to the storage capacity of the ice storage itself and α_{ST} ($/kWh) was the cost coefficient furnished by the manufacturer. Other works preferred the use of semi-empirical formulas. Towler and Sinnot [6] proposed an easily adaptable cost estimation relation for storage tanks displayed in Equation (5.2), as follows:

$$Z_{Tank} = a + bV_{ST}^n$$

	a	b	n
Floating roof	113,000	3,250	0.65
Cone roof	5,800	1,600	0.7

where

$$\tag{5.2}$$

V_{ST} (m^3) refers to the volume of the tank, and a, b, and n are coefficients depending on the type of tank as detailed above. Sanaye and Shirazi [3] and Shao *et al.* [7] both approximated ice storage tank cost using the formula

$$Z_{ST} = 8.67 \times 10^{2.9211} \exp(0.1416 \log V_{ST}) \tag{5.3}$$

Rathgeber *et al.* [8,9] presented an insightful study on the framing of TES economic analyses using two separate approaches, namely, a top-down approach and a bottom-up approach. The key concept introduced is the segmentation of TES installations into multiple user classes, namely, "industry", "building", and "enthusiast". These distinctions provide useful estimations for key parameters that impact the overall cost of the TES system, such as the interest rates and storage capacity.

The top-down approach determines the maximum acceptable storage capacity cost (SCC_{acc}) by assuming that the cost of energy supplied by the storage does not exceed the cost of energy from the market. This can be computed as follows:

$$SCC_{acc} = \frac{(REC \cdot N_{cycles})}{ANF} \tag{5.4}$$

$$ANF = \frac{(i+1)^n \cdot i}{(i+1)^n - 1} \tag{5.5}$$

where ANF refers to the present value annuity factor, REC ($/kWh) refers to the reference energy costs, and N_{cycles} refers to the number of charge–discharge cycles in a year. i refers to the interest rate and n refers to the payback period.

Based on the assumed user classes as elaborated earlier, it is assumed (based on known case studies and the literature) that the interest rates for the Industrial, Building, and Enthusiast tiers are >10%, 5%, and 1%, respectively. Accordingly, the respective payback periods are assumed to be 5 years, 15–20 years, and 25 years. The resulting ANF for each user class can then be plotted against varying payback periods as portrayed in Figure 5.5.

An acceptable ANF of 0.25–0.30 (~5 years payback) for the Industrial sector is apparent in Figure 5.5, while the Building sector permits an ANF spanning 0.07–0.10 for longer payback periods. Lastly, the Enthusiast class has low ANFs ranging from 0.04 to 0.06. A similar expected range of reference energy costs is also employed to evaluate the TES system costs. The economic boundary conditions of the three user classes for both ANF and REC are displayed in Table 5.1.

For analysis of the high-cost and low-cost scenarios for each user class, the high-cost case considers the maximum REC and lowest ANF, while the low-cost case is the converse. Rathgeber *et al.* [8,9] further state that the operational costs should be accounted for when they are comparable to the capital costs, such as in the case of mobile TES systems.

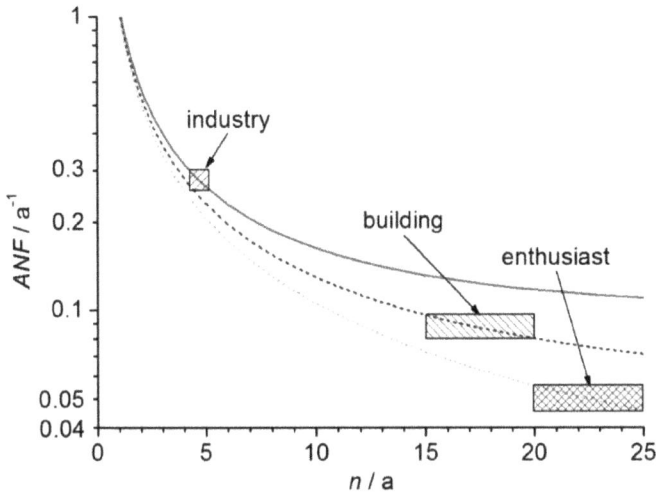

Figure 5.5. ANF against payback period for each user class as presented by Rathgeber *et al.* [8,9]. The framed regions demarcate the acceptable ANF for each use class.

Table 5.1. Economic boundary conditions of the three user classes for the top-down evaluation [8,9].

User class	REC/($/kWh)	ANF
Industrial	0.02–0.04	0.25–0.30
Building	0.06–0.10	0.07–0.10
Enthusiast	0.12–0.16	0.04–0.06

In the bottom-up approach, Rathgeber *et al.* [8,9] utilize data from several TES case studies and their respective costs. When all the operational details and parameters, as well as all component-level costs, are investigated, a comparison can be made between the real storage capacity costs (SCC_{real}) and the SCC_{acc} trends as highlighted in the top-down approach. The SCC_{acc} trends against N_{cycles} of the three user classes are computed using Equation (5.3) and are as shown in Figure 5.6.

It is worth noting that for a fixed set of charge–discharge cycles in a year, the SCC_{acc} of a TES system depends solely on the economic environment of the user. Additionally, the difference between the costs of the enthusiast and industrial user classes is also highly

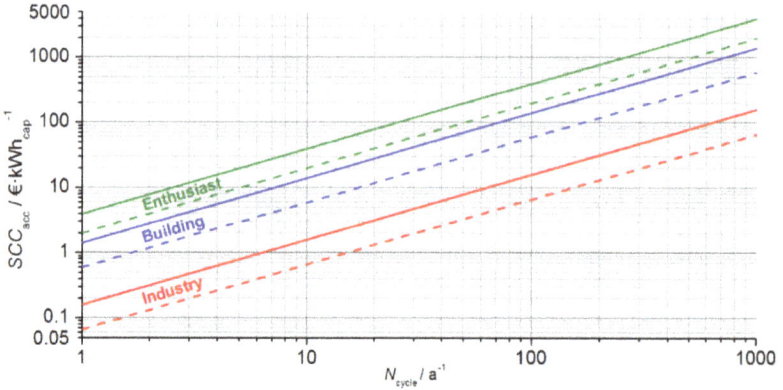

Figure 5.6. SCC_{acc} trends against N_{cycles} for each user class. The solid lines indicate the high-cost case, while the dashed lines indicate the low-cost case for each respective user class [8,9].

evident. A factor of almost 60 is observed between the industrial class low-cost case and enthusiast class high-cost case. The increase of the SCC_{acc} for higher storage cycles highlights the impact of larger energy turnovers on the system.

The value of SCC_{real} can be defined simply by the total investment cost divided by the storage capacity, as follows:

$$SCC_{real} = \frac{Z_{total}}{SC} \qquad (5.6)$$

where Z_{total} ($) refers to the total investment cost and SC (kWh) refers to the storage capacity of the TES system. The analysis of the SCC_{real} is divided into long-term and short-term storage for ease of analysis. Figure 5.7 shows the comparison exercise conducted between the SCC_{acc} and SCC_{real} values.

From the 26 TES systems reviewed by Rathgeber *et al.* [8], a few key observations were made (currency in €) as follows:

(1) The storage capacity cost limits for seasonal TES systems (long-term storage) operating up to a maximum of 2 cycles/year was € 3/kWh for the building user class and € 0.4/kWh for the industrial user class. Therefore, seasonal TES systems are observed to be economically viable for very large capacity hot water sensible TES (STES) systems.

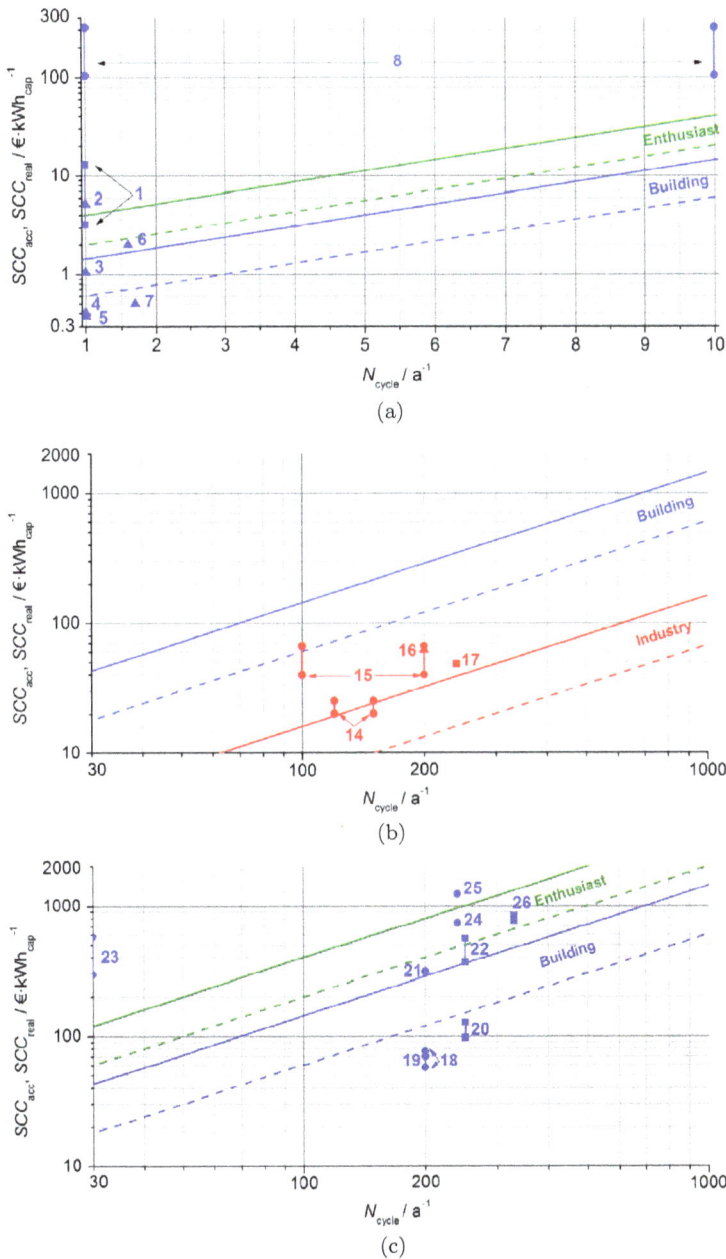

Figure 5.7.   Comparison between SCC_{acc} and SCC_{real} for (a) long-term storage cases, (b) short-term storage cases for industrial class, and (c) short-term storage cases for enthusiast and building user classes [8].

(2) Despite hot water storages up to 30 m^3 in storage volume being more applicable for the building user class, they can still be economically viable at the industrial level if the number of storage cycles is sufficiently high.

(3) Ice-based LTES systems are found to be cost-effective for short-term storage at the industrial scale.

In breaking down the key contributing cost factors for the cases reviewed through the bottom-up approach, Rathgeber *et al.* [8] observed that the charge/discharge system as well as the storage enclosure account for far higher shares of the overall costs compared to the thermal storage material used. This finding implies that developing more cost-effective methods of storage container construction/design as well as charge/discharge devices are key to economic viability for TES implementation.

5.1.3. *TES-integrated applications and processes*

A major part of the process integration into TES involves district thermal systems that cater to either urban heating or cooling demands. Given the rising global temperatures, more attention has been paid to cooling applications, particularly for air-conditioning and mechanical ventilation, which account for a high percentage of energy demand in tropical climates [10], as illustrated in Figure 5.8.

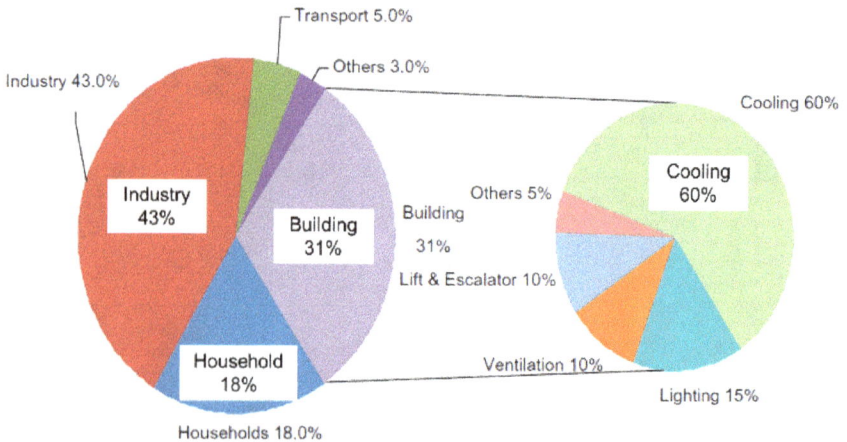

Figure 5.8.   Distribution of electrical use in Singapore for the Building sector [10].

Guelpa and Verda [11] posit that TES systems bring about numerous benefits for district thermal systems such as increased operational flexibility as well as offsetting large supply and demand variations; leading to improved overall system performance. Additionally, TES systems enable the integration of multi-energy systems such as cogeneration, solar, or even water treatment — promoting the economization of resources such as waste heat or cooling. Guelpa and Verda [11] have judiciously classified TES integration into district thermal systems as portrayed in Figure 5.9.

Besides the usual material-based classification of TES (sensible, latent, and thermochemical), the duration of storage is also a key distinguishing factor between TES systems. Long-term storages are often used for seasonal TES where the heat during summer is used for winter heating, while the cold during winter is stored for summer cooling. Examples of these include Aquifers and Borehole TES systems, which have been briefly described in Chapter 1. Short-term storages involve high charge–discharge rates for daily consumption and are often deployed as chilled water/hot water storage tanks at building and industrial levels. The last classification type is based on its distribution. TES can be localized in a central location or distributed to individual suppliers/users within the network, or in

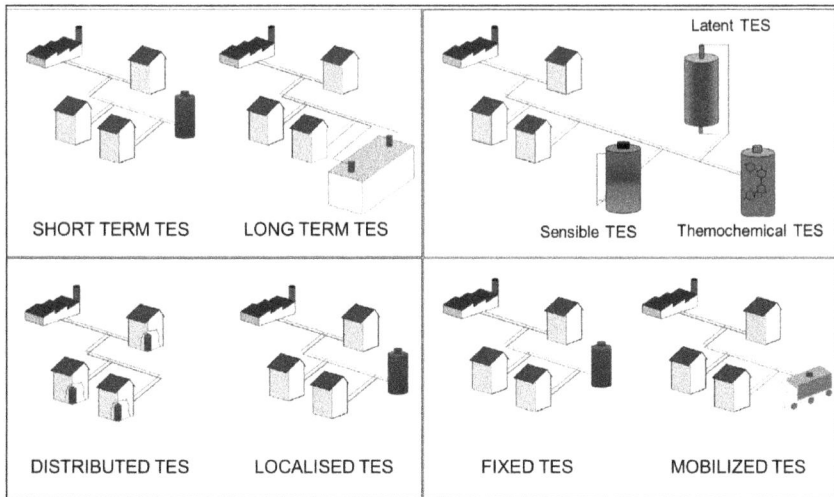

Figure 5.9.   Classification of TES-integrated district thermal systems by Guelpa and Verda [11].

some niche cases even mobilized. Mobilized TES systems involve movable TES units which are transported between source and user using trucks.

Zhang *et al.* [12] studied the integration of an ice storage system into a district cooling system (DCS), as portrayed in Figure 5.10(a). They developed a load-response model using *TRNSYS* for real-time demand change response. It was found that the ice storage-integrated DCS model increased electrical consumption by 9% but enhanced the overall cooling capacity of the system by 20%. As illustrated in Figure 5.10(b), the storage was found to reduce the electrical consumption from the power grid as well as trimmed operating costs. In sum, DCS operating costs were reduced by 6.78% and the overall system efficiency was improved by 9.46%.

Numerous works carried out by Anderson *et al.* [13] and Cox *et al.* [14] have also explored the control and operation of TES-integrated district thermal systems. Anderson *et al.* [13] examined a campuswide district heating and cooling system (DHCS) at the University of Idaho using the *TRNSYS* platform. The DHCS comprised both coal and biomass burning for heat generation. Biomass burning provided up to 95% of the process steam utilized by the university. During the warm seasons, the produced steam is fed to an absorption chiller to provide daytime cooling. Electrical chillers are operated during the night to charge the cold TES, shown in Figure 5.11(a). The stored cold energy is then discharged during the daytime. When a simple operating schedule is implemented, it was observed that the electrical consumption was reduced by 38% as revealed in Figure 5.11(b). Concurrently, CO_2 emissions were observed to reduce by 428,800 kg during the cooling season. This corresponded to an overall operating cost reduction of $140,000 — a significant amount.

As TES systems have evolved to become a mature technology in district thermal systems, much of the work today is focused on optimizing their energy use and cost by employing novel operating strategies. These will be covered in the ensuing section.

Recently, one of the more niche forms of cooling supply and distribution that have attracted much academic interest has been the cold energy recovery from LNG regasification. LNG is often stored within coastal LNG terminals at cryogenic temperatures of around –163°C and under atmospheric pressure conditions. Conventional methods

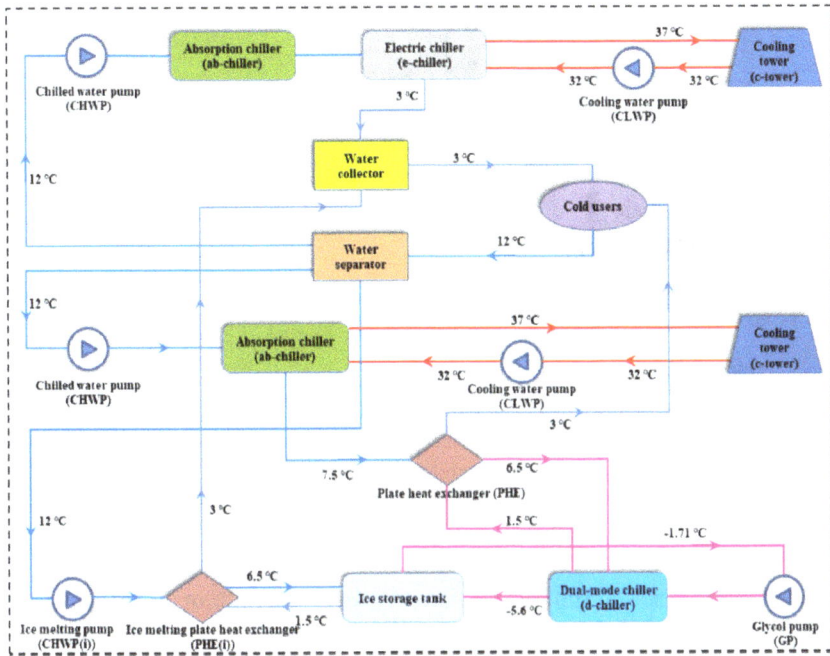

(a)

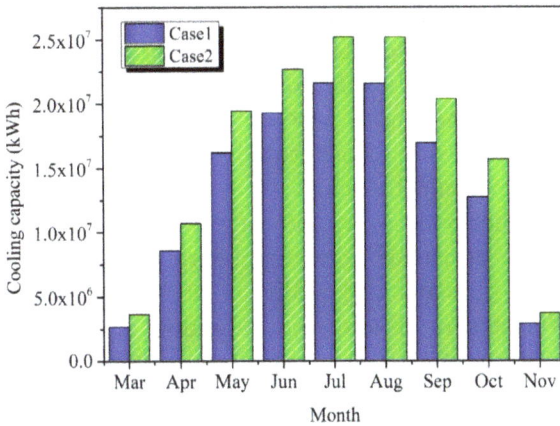

(b)

Figure 5.10.   (a) Schematic representation of the ice storage integrated with DCS and (b) the improved cooling capacity of the system integrated with ice storage in comparison to the baseline case. Case 1 refers to the baseline, while Case 2 refers to the ice storage-integrated DCS [12].

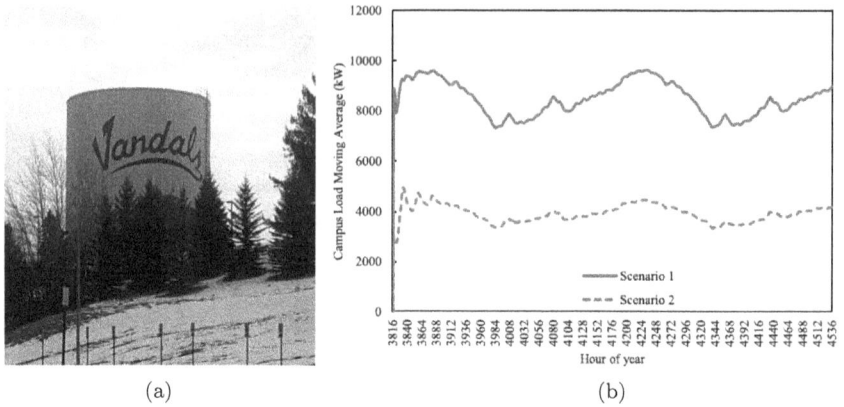

(a)   (b)

Figure 5.11. (a) The cold TES integrated with the campuswide DHCS at the University of Idaho and (b) the improvement in overall power consumption under the proposed operating strategy as studied by Anderson *et al.* [13].

of regasifying this LNG to gaseous NG for combustion in combined heat and power (CHP) plants involve heat exchange with seawater flow in Open Rack Vaporizers (ORVs) [15]. These methods have led to a large amount of waste of cold exergy and energy to the environment being recovered. One of the earliest adopters of cold-recovery processes for LNG regasification was the Osaka Gas company in the early 1970s. It utilized the LNG cold for air-conditioning, power generation, and carbon dioxide liquefaction in its *Senboku* LNG Terminal [16]. Over time, more processes such as air liquefaction and ethylene production were incorporated into terminals until 100% cold energy utilization could be achieved, as portrayed in Figure 5.12.

While cold energy recovery is a key focus of this aspect of TES research, a more pertinent performance indicator is the cold exergy efficiency, which indicates the quality of cold that could be recovered from the LNG stream. The exergy of a flow stream can be simply described as

$$\phi = (H - H_o) - T_o(S - S_o) \tag{5.7}$$

where φ (kW) refers to the exergy of the flow stream, and H (kW) and S (kW/K) refer to the enthalpy and entropy of the flow stream, respectively. The subscript o refers to the reference base state, which is often taken to be the ambient condition, and T_o (K) refers to the reference condition temperature.

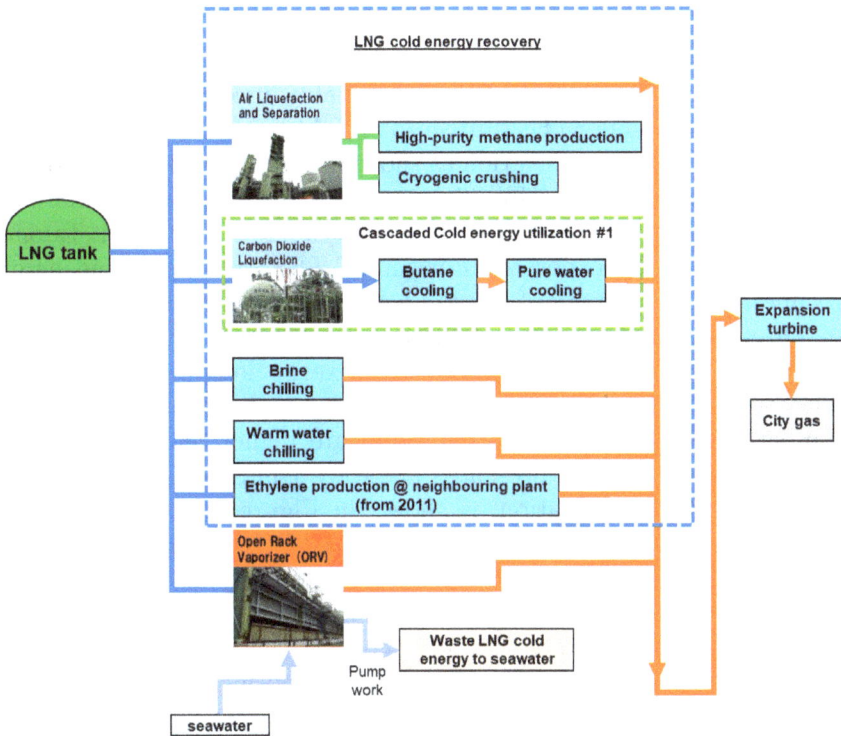

Figure 5.12. Outline of the cold recovery processes at Senboku Terminal 1 by the Osaka Gas Company [16].

The exergy efficiency of the process is then defined as the ratio of the utilized exergy to the exergy expended by the LNG through the regasification process or processes [16], as follows:

$$\eta_\phi = \frac{\phi_{\text{process}}}{\phi_{\text{LNG,in}} - \phi_{\text{LNG,out}}} \qquad (5.8)$$

where the subscripts LNG, in and LNG, out refer to the inlet and outlet conditions of the LNG stream, respectively. φ_{process} refers to the exergy utilized by the process. Evaluation of the Senboku Terminal 1 by Yamamoto *et al.* [16] revealed that an improvement in the exergy efficiency from 45% to 54% can be achieved after the incorporation of the ethylene plant.

Khor *et al.* [17] noted that research works that were centred on cold recovery from LNG regasification disproportionately focused on cryogenic power generation and liquid air energy storage (LAES). Both of these applications see low industrial uptake due to scalability issues and cost. Peng *et al.* [18] notably incorporated multiple pressurized-propane STES units into an LAES system that was integrated with LNG regasification as shown in Figure 5.13. It was found that up to 88% round-trip efficiency could be achieved; a marked improvement over the 60% baseline. Ryu *et al.* [19] and Tafone *et al.* [20] both incorporated cryogenic LTES systems as means of recuperating cold energy from the LAES during discharge.

Direct thermal transfer methods for cold recovery do not seem to be popular due to excessive cold exergy losses. Nevertheless, TES systems, particularly LTES systems, can be deployed for low-temperature cryogenic storage to facilitate higher retention of cold exergy. Khor *et al.* [17] presented a 4-stage LTES cold-recovery system for LNG regasification as shown in Figure 5.14. A mobilized concept was also featured where the charged PCMs were transported by trucks to the end users. The respective PCM phase-change temperatures for each consecutive stage were –86°C, –67°C, –33°C, and –3°C as displayed in Table 5.2. For the district cooling end users, an overall exergy efficiency of 1.77% was deduced for the system while the deep-freezing application realized a higher overall performance of 15.96%. Among the end-user applications tested, the 4-stage LTES concept performed marginally better (11.08%) than the thermal network concept proposed (10.04%), but it was still comparatively lower than the baseline case of 13.02%.

Shao *et al.* [7] proposed a multi-stage quad-generation concept that integrates cryogenic LTES with a carbon dioxide power cycle, chilled water cooling, and an air cooling/dehumidification process. As depicted in Figure 5.15, the LTES stage constitutes the first stage of the system and is primarily responsible for the low-temperature pre-heating stage for the LNG during its saturated liquid phase. The PCM employed in the LTES had a phase-change temperature of –11°C and a latent heat of fusion of –301 kJ/kg [21]. The second stage comprises the power cycle stage, involving a CO_2 Rankine cycle driven by waste heat from the CHP process on the evaporator side and cooled by the regasifiying LNG on the condenser side. The third and fourth stages of the process recover the remaining cold from the regasified LNG (now in its gaseous phase). Direct thermal transfer occurs between returning chilled water (at 12°C) and ambient air to

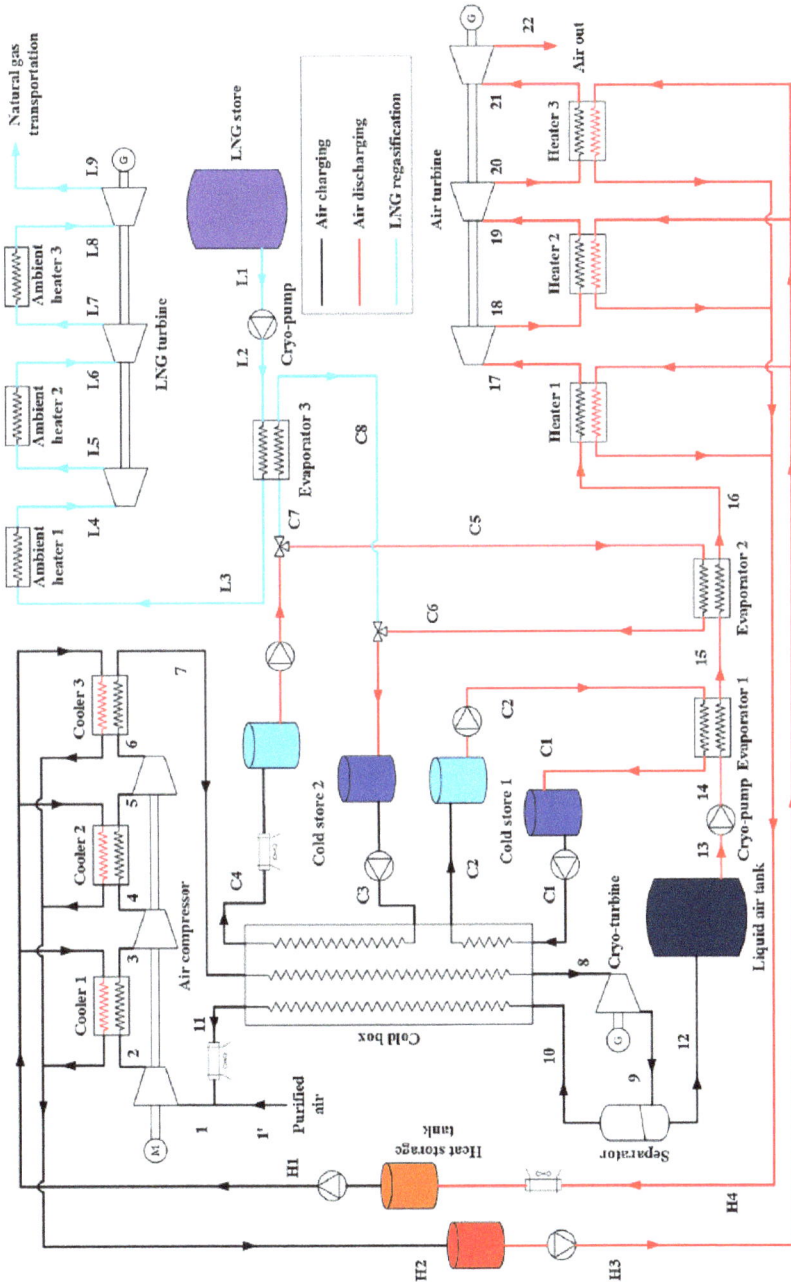

Figure 5.13. LAES system integrated with LNG regasification and multiple propane-based cold TES units by Peng *et al.* [18].

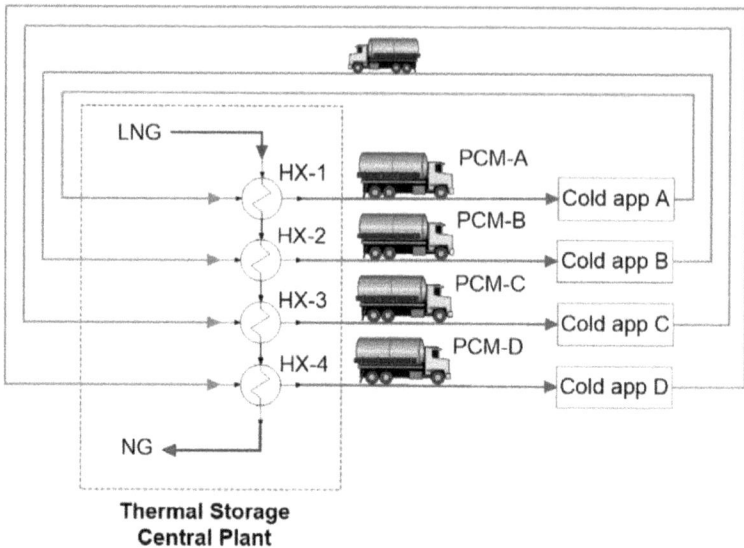

Figure 5.14.   4-Stage LTES for cold recovery from LNG regasification as assessed by Khor *et al.* [17].

Table 5.2.   PCM properties for the 4-stage LTES cold recovery system as studied by Khor *et al.* [17].

	PCM-A	**PCM-B**	**PCM-C**	**PCM-D**
Name/Composition	24.8 wt% HCl	24.0 wt% LiCl	SN33 [21]	AN03 [21]
Phase-change temperature	−86°C	−67°C	−33°C	−3°C
HTF flow rate	1.7 kg/s	2.2 kg/s	12.1 kg/s	3.1 kg/s
LNG flow rate	7.4 kg/s			

produce cooling, dehumidification, and potable water from dew-point condensation.

The overall process was thermodynamically modelled and validated with a lab-scale tri-generation system which utilized liquid nitrogen in place of LNG due to safety concerns. The LTES capacity was sized according to the rate of LNG heat gain across the LTES stage multiplied by the charging time. Based on the PCM properties and the PCM volume fraction within the LTES, the volume of the

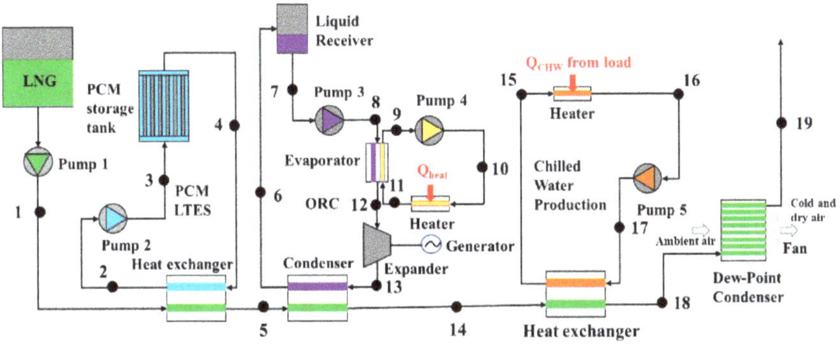

Figure 5.15. Schematic representations of (a) the 4-stage quad-gen cold-recovery system and (b) the energy conversion process by Shao *et al.* [7].

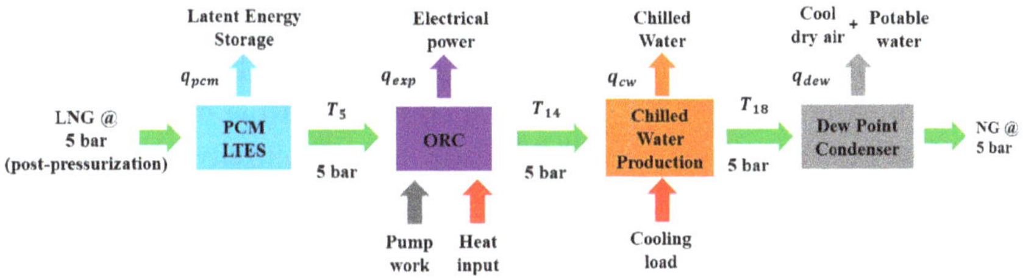

LTES is computed as

$$Q_{\text{LTES}} = m_{\text{lng}} (h_{\text{out}} - h_{\text{in}}) \tag{5.9}$$

$$E_{\text{TES}} = (Q_{\text{LTES}} - Q_{\text{loss}}) \times t_{\text{charge}} \tag{5.10}$$

$$V_{\text{TES}} = \frac{E_{\text{TES}}}{\varepsilon \rho_{\text{pcm}} [c_{f,\text{pcm}} (T_{\text{hot}} - T_{fc,\text{pcm}}) + c_{s,\text{pcm}} (T_{fc,\text{pcm}} - T_{\text{cold}}) + h_L] + (1 - \varepsilon) \rho_{\text{htf}} c_{\text{htf}} (T_{\text{hot}} - T_{\text{cold}})} \tag{5.11}$$

where Q_{LTES} and Q_{loss} (kW) refer to the LTES charging rate and the heat ingress rate to the LTES during the charging period, t_{charge} (s), respectively. m_{lng} (kg/s) refers to the mass flow rate of the LNG stream. h_{out} and h_{in} (kJ/kg) refer to the exiting and entering enthalpies of the LNG, respectively. E_{TES} (kJ) and V_{TES} (m^3) refer to the storage capacity and storage volume of the LTES, respectively.

T_{hot} and T_{cold} (°C) refer to the respective discharge and charge temperatures of the LTES, respectively. ρ (kg/m^3), ε, h_L (kJ/kg), and c (kJ/kg · K) refer to the density, PCM volume fraction, latent heat of fusion, and specific heat, while the subscripts s, pcm, f, pcm, and htf refer to the solid-phase PCM, liquid-phase PCM, and the HTF of the LTES, respectively.

A multi-objective optimization via genetic algorithm (GA) was carried out to obtain the optimal design conditions for the process at multiple levels of LNG demand. Seven decision variables, including the entering and exiting temperatures of each stage, as well as the LTES hot and cold side temperatures, were chosen for optimization as shown in Figure 5.16. The objective functions included the exergy efficiency for maximization, system cost for minimization, and CO_2 emissions for minimization. At the 5 kg/s LNG demand, it was found that the exergy efficiency obtained from a triple-objective optimization process reaped a 12.3% improvement compared to the best result from the single-objective optimization. The Sankey chart for the triple-objective optimized case is as portrayed in Figure 5.17. Key findings, however, revealed that, despite the comparable performances between this work and that of Khor *et al.* [17], much more exploration into the integrating effectiveness of LTES during LNG regasification still has to be conducted to realize a more effective method to design and operate the overall system.

5.2.   Control and Monitoring for TES-Integrated Processes

As highlighted in Chapter 1, a key focus in the integration of TES and associated processes today, namely, district thermal systems, is the operating strategy and control methods. Cox *et al.* [14] opine that a "smart" grid requires the use of autonomous controllers to manage these systems so that the "smart" grid is able to respond quickly to changes in environmental and economic conditions with minimal human intervention. To do so, control strategies that allow the system and processes to select their operating points to achieve the most optimal outcomes are critical for users and operators.

As novel AI-based controls and metaheuristic optimization methods are becoming more prevalent in industrial and pilot settings,

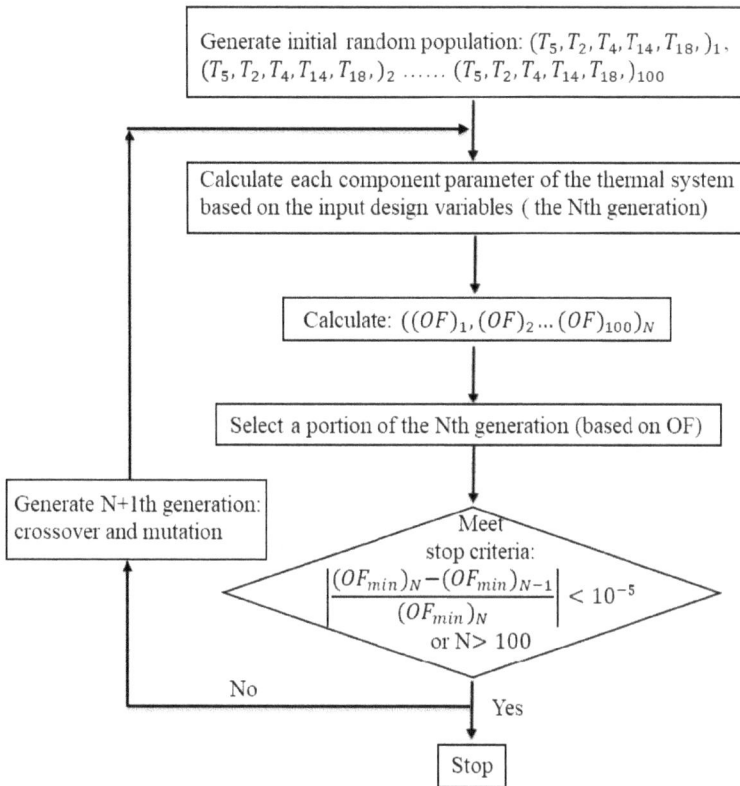

Figure 5.16. Genetic algorithm (GA) method used for the optimization of the quad-gen system proposed by Shao *et al.* [7].

much research work has been dedicated to examining the impacts of these control strategies on actual installations under living-lab and simulated conditions.

5.2.1. *Operating strategies*

As previously elaborated in Chapter 1, district thermal systems, in particular DCS with integrated TES, tend to operate based on either full-storage, demand-limiting, or load-levelling schemes, as shown in Figure 5.18. Numerous scenarios in the literature have shown a tendency to prefer optimizing the system design for load levelling, wherein the chiller load is kept constant throughout the day. This is executed by sizing the chiller and TES such that the total charging

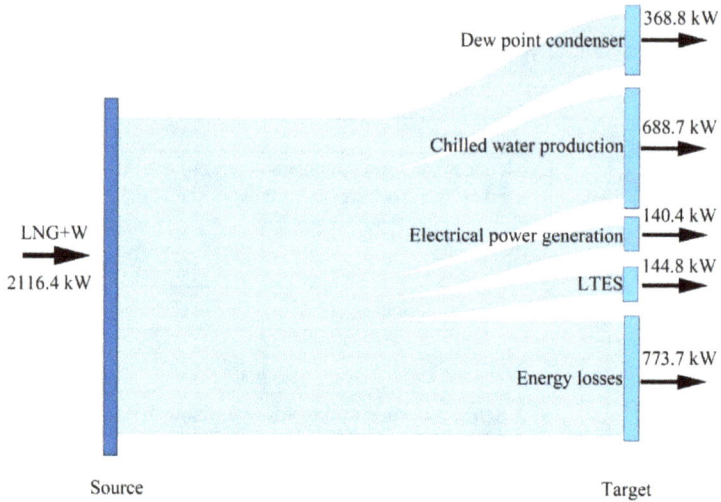

Figure 5.17.   Sankey chart for the triple-objective optimized design by Shao *et al.* [7].

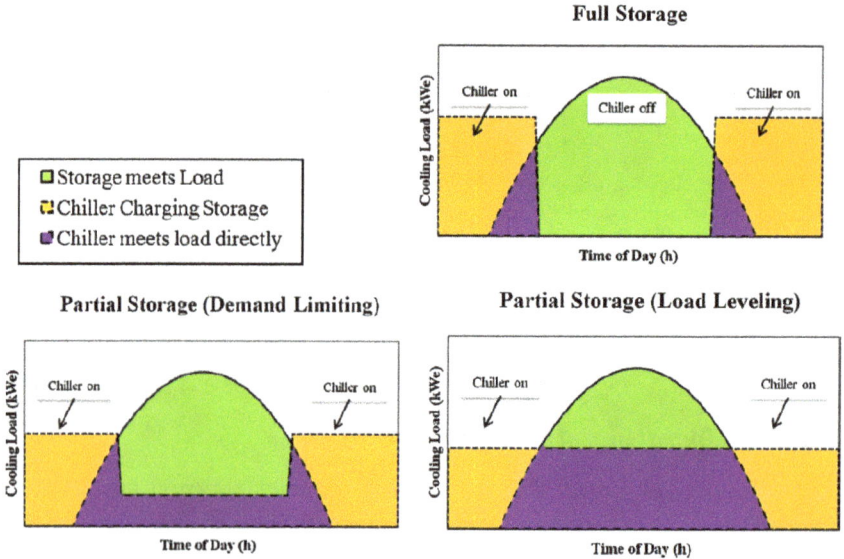

Figure 5.18.   Full-storage, demand-limiting, and load-levelling operating schemes for TES-DCS systems [22].

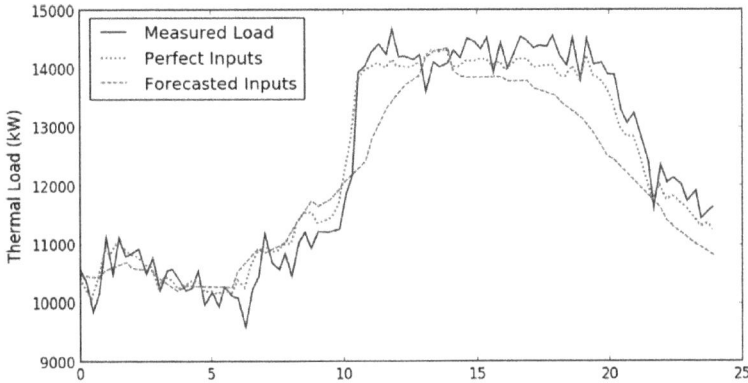

Figure 5.19.   Measured and predicted cooling load by Cox *et al.* [14].

capacity during the off-peak/low-demand hours is equalled to the excess load beyond that of the chiller operating load during peak hours. This can be simply described as

$$E_{\text{TES}} = \int_0^{t_{\text{offpeak}}} Q_{\text{chiller}}\, dt = \int_0^{t_{\text{peak}}} Q_{\text{peak}} - Q_{\text{chiller}} dt \quad (5.12)$$

Accordingly, the demand-limiting scheme is simply described by equating the excess load during peak periods to the charged TES capacity during off-peak periods. For the full-storage scheme, the chiller is assumed to be non-operating during peak hours and hence the charged capacity during off-peak periods is equivalent to the full-discharge capacity of cooling during peak hours.

These generalized sizing rules and methods are typically idealized as they do not take into account the daily variations in the environment and users' conditions as highlighted in Figure 5.19.

Additionally, fluctuations in peak and off-peak electrical tariffs may render preference for another operating strategy from an economic standpoint over time. These variabilities emphasize the importance of the implemented control strategy to optimize real-time conditions as well as the shifting priorities of users and stakeholders.

5.2.2. *"Smart" control systems*

The term "smart systems" is often used liberally with regards to varying manifestations of control systems. Technically, it is attributed to devices or systems that [23]

(1) Employ integrated sensing capabilities to capture environmental changes or changes in environmental data;
(2) Process data independently via in-built software and electronics; and
(3) React to key information by means of actuation.

Østergaard *et al.* [24] distinguish future "4G" or fourth-generation district cooling (4GDC) systems from their predecessors by the manner in which such "smart" systems managed to improve the operating efficiencies of DCS. This is primarily due to the integration potential of multiple energy networks including renewable energy and waste heat sources as portrayed in Figure 5.20.

Ironically, within the decentralized and integrated nature of "smart" 4GDC systems, individual technologies in the network may end up performing less efficiently but result in the overall system performing more efficiently. In contrast, the DCS of the past were focused on varying iterations of thermal distribution network design. For instance, Østergaard *et al.* [24] attribute much of the progress in 3rd generation (3G) DCS to the advancement of TES systems.

In summary, the prime motivation for 3GDC systems is to maximize the utility of all energy resources available and phase-out refrigerants with high ozone-depleting potential (ODP). In contrast, 4GDC systems aim to synergize combined heating and cooling (CHC) systems. In this section, a few examples of control strategies will be reviewed to present recent developments in this field of study.

Anderson *et al.* [13] observed in their study that only a simple change to the operating schedule was required to optimize the energy distribution during the cooling season. It was also found that the operating strategy of the DCS involved large amounts of exergy loss as shown in Figure 5.21. This was attributed to the chillers operating continuously while charging a fully charged TES during the nighttime period.

Energy generation systems Energy distribution and storage systems End users

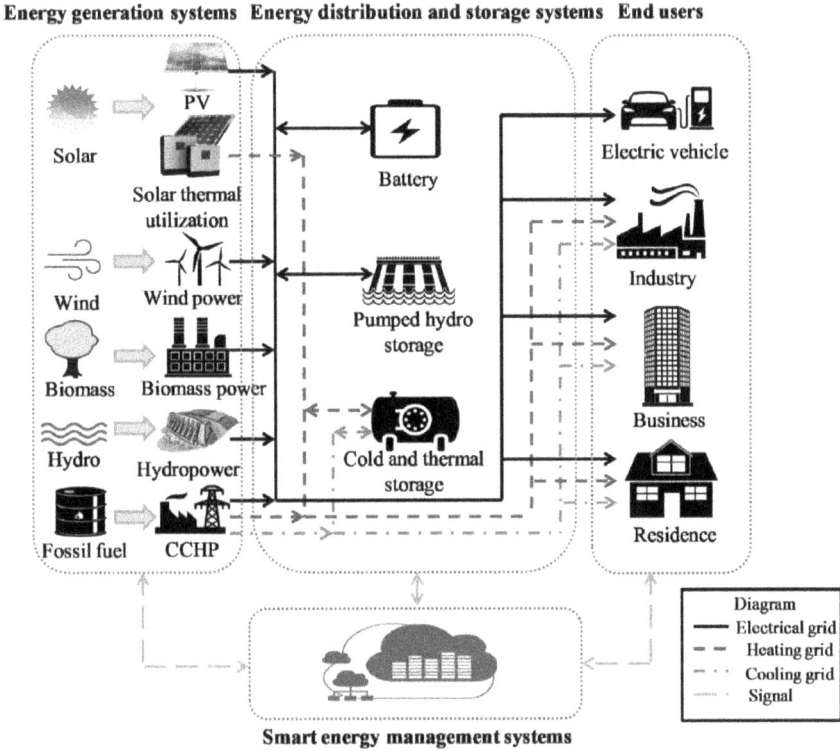

Figure 5.20. A "smart" energy system as conceived by Xu *et al.* [25].

The exergy balance equation for the system could be described comprehensively as

$$\left(\sum \phi_{Q_{\text{in}}} - \sum \phi_{Q_{\text{out}}}\right) + \left(\sum \phi_{W_{\text{in}}} - \sum \phi_{W_{\text{out}}}\right)$$
$$+ \left(\sum \phi_{f_{\text{in}}} - \sum \phi_{f_{\text{out}}}\right) - \phi_{\text{des}} = \frac{d\phi_{\text{des}}}{dt} \qquad (5.13)$$

where $\phi_{Q_{\text{in}}}$ and $\phi_{Q_{\text{out}}}$ (kW) refer, respectively, to the incoming and outgoing exergy rates from the heat transfer across the system boundary. $\phi_{W_{\text{in}}}$ and $\phi_{W_{\text{out}}}$ (kW) refer to the exergy rates of work in and out of the system, and $\phi_{f_{\text{in}}}$ and $\phi_{f_{\text{out}}}$ (kW) refer to the specific exergy rates of the fluid mass flow rates in and out of the system, respectively. ϕ_{des} refers to the exergy destruction rate. Without seeking to perform any equipment alteration, a fixed change in the

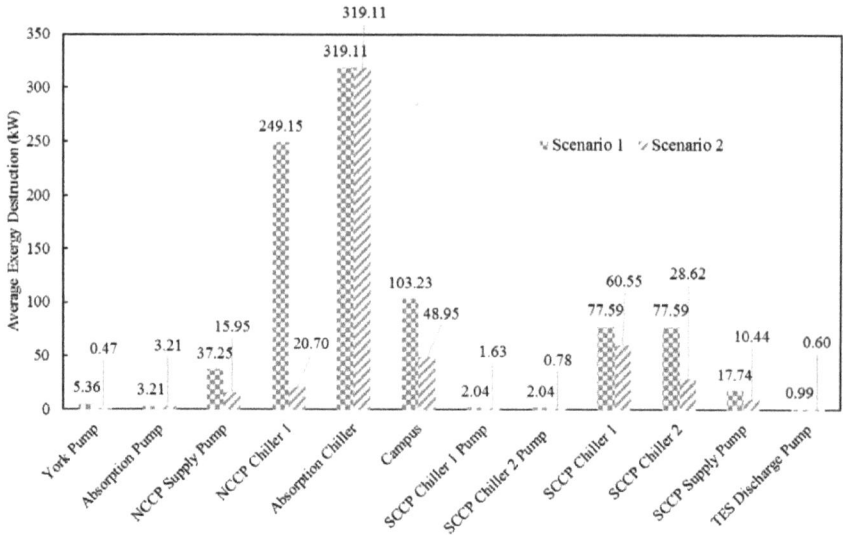

Figure 5.21.   Computed exergy destruction breakdown by components for June 2016 [13].

Table 5.3.   Proposed operating scheduling changes by Anderson *et al.* [13].

	Absorption chiller	Chiller 1 (North campus)	Chiller 1 (South campus)	Chiller 2 (South campus)
Baseline	Always on	Always on	Always on	Always on
Proposed	Always on	15:00–17:00	01:00–02:00	10:00–19:00

operating schedule was proposed instead and tested on TRNSYS as shown in Table 5.3.

Savings in cost, electrical consumption, and CO_2 emissions were reported from the proposed scheduling as detailed in Table 5.4. Based on Idaho's electrical tariff of $0.059/kWh, a $35,000 cost reduction for the month of June was achieved compared to the baseline. Total electrical consumption was also found to be reduced by 592,500 kWh and an overall emissions reduction of 107,200 kg was computed. Over

Table 5.4. Improvements arising from the scheduling changes by Anderson *et al.* [13].

	Electricity (kWh)	Cost ($)	CO_2 (kg)
Baseline	945,700	55,800	171,200
Proposed	353,500	20,860	64,000
Savings			
June	592,200	35,000	107,200
4-Month period	2,368,800	140,000	428,800

the entire cooling period of four months, a cost savings of up to $140,000 was estimated.

While simple changes can lead to improved outcomes in some cases, others require a higher level of real-time control due to additional complexities in the system. Cox *et al.* [14] developed an elaborate control strategy incorporating Neural Network (NN) coupled with a Genetic Algorithm (GA) to maximize cost savings. The case study involved a DCS operating at Mississippi State University (MSU) which employed an extensive ice storage system as shown in Figure 5.22.

The overall methodology employed by Cox *et al.* can be divided into three key parts: load prediction, system modelling, and optimization and control. The first part involves the control system estimating the expected system load for a specific time period based on measured environmental parameters: dry-bulb temperature (T_{db}), relative humidity (RH), and global horizontal irradiation (GHI). In this study, a nonlinear autoregressive with exogenous inputs (NARX) neural network (NN) was developed to predict the load. The model was trained using hourly weather forecast data for a 5-day period. The NN architecture was earlier illustrated in Figure 1.30. Each exogenous input vector for the NN is described as

$$x = [T_{db} \, \text{RH} \, \text{GHI} \, h_0, \ldots, h_n \, y_{t-1}, \ldots, y_{t-n}] \qquad (5.14)$$

where h_o to h_n refer to the hours of the day, with n being the nth input neuron indicating which neuron is active for that time of the day, and y refers to the predicted cooling load for that point.

Following the load prediction step, the DCS is then modelled. A physics-based model is first constructed, employing fundamental

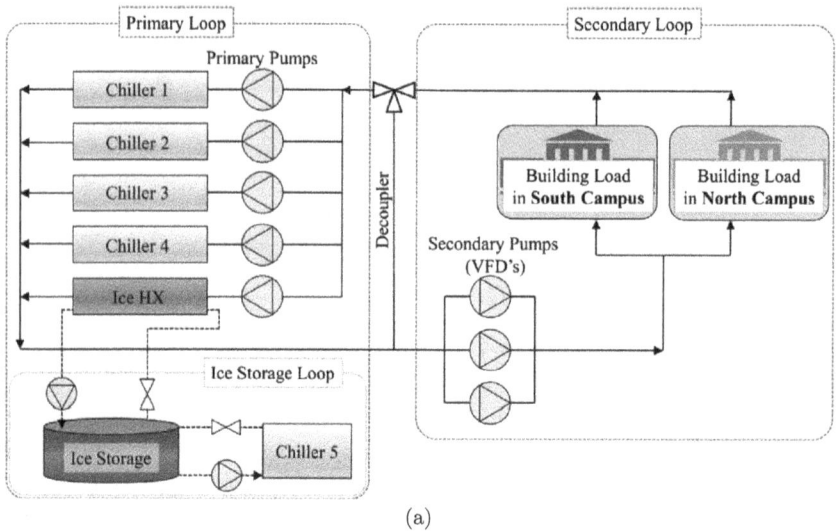

(a)

(b)

Figure 5.22.   (a) Schematic representation of the DCS and (b) the ice storage tank system at MSU [14].

thermodynamic relations involving mass flow rates, temperatures, and specific heats of the materials used. The chiller plant controls are then included to ensure that the model closely resembles that of the actual scenario. Next, validation of the model is performed as depicted in Figure 5.19. A neural network model was also developed to accelerate the system development time and generalize complicated plant designs. Accordingly, the response time of the proposed model-predictive control (MPC) is improved. The validation between

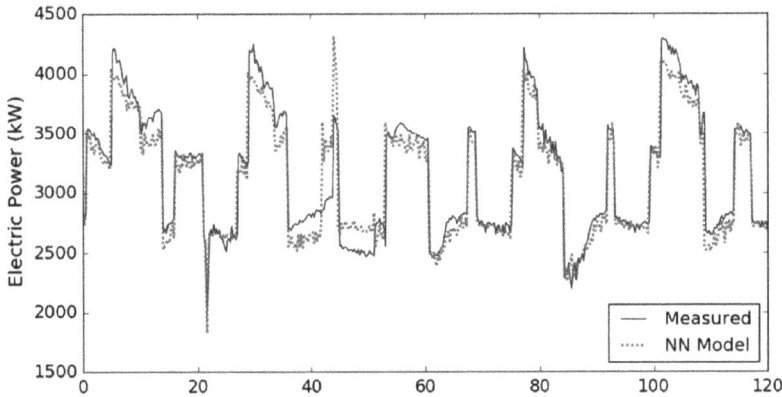

Figure 5.23. Comparison between the NN model and the measured data by Cox *et al.* [14].

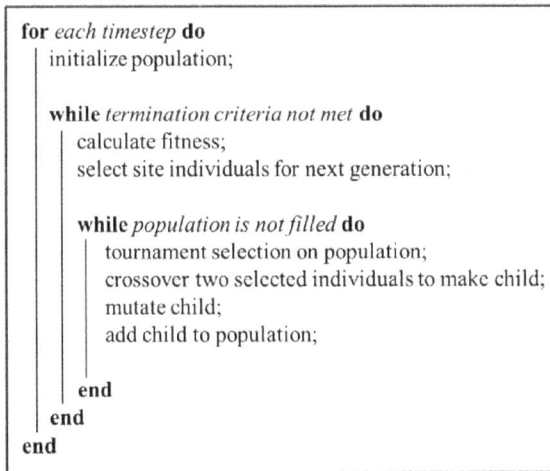

Figure 5.24. The genetic algorithm (GA) process [14].

the predicted electrical consumption from the NN model and the measured data is demonstrated in Figure 5.23.

Once the NN model was constructed, an optimization scheme had to be put in place to determine the optimal operating setpoints. A genetic algorithm (GA) was selected for this task, as it is naturally binary-based, which suits the on/off nature of the plant controls. Additionally, the chromosomes in the algorithm could be well-suited

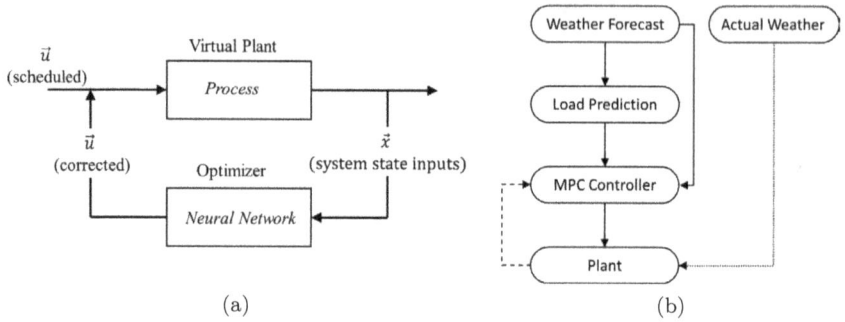

Figure 5.25. (a) The block diagram for the MPC and (b) data flow of the process [14]. u refers to the control signal and x refers to the system state inputs.

for this problem type. In a similar vein, Shao *et al.* [7], as was shown earlier in Figure 5.16, proposed that a genetic algorithm is capable of replicating the "survival of the fittest" evolutionary process by tournament selection of a specific group of candidate variables. The algorithm is depicted in Figure 5.24.

The GA was considered to have converged if the optimal solution did not change for 30 generations [14]. Figure 5.25 outlines the structure of the MPC as well as the data flow of the control algorithm.

The MPC was utilized in-line so that its ongoing states were updated at every time step. The formulation of the NN model is considered a black-box approach and only on–off signals are provided to the chiller and ice storage system, with the external variables as inputs. This can be described simply as

$$\left[P_e, \dot{Q}_{ice}\right] = f(\overrightarrow{u_1}, \overrightarrow{u_2}, \overrightarrow{u_3}, \overrightarrow{u_4}, \overrightarrow{u_c}, \overrightarrow{u_d}) \tag{5.15}$$

where $\vec{u}_i$ refers to the control signal to the chiller i, $\vec{u}_c$ is the control signal for ice storage charge, and $\vec{u}_d$ is the control signal for ice storage discharge. The part load ratio (PLR) of the chiller plant is approximated by assuming the secondary load (Q_{sec}) to be equivalent to the primary load (Q_{pri}), as follows:

$$\text{PLR}(t) = \frac{Q_{pri}(t)}{Q_{max}(t)} \tag{5.16}$$

$$Q_{max} = u_1 Q_{max,ch1} + u_2 Q_{max,ch2} + u_3 Q_{max,ch3}$$
$$+ u_4 Q_{max,ch4} + u_d Q_{max,ice} \tag{5.17}$$

where $Q_{max}(t)$ refers to the maximum chiller plant loading at time step t and the subscripts ch1, ch2, ch3, and ch4 refer to the 4 chillers, respectively. A chiller is added when the PLR exceeds 0.95 and subtracted when PLR dips below 0.4. Consequently, the model is simplified into a function with initial conditions that are easily optimized considering the charging and discharging statuses. The objective function for the optimization is thus formulated as

$$J = \int_{t_0}^{t_f} P\left(u\left(t\right)\right) \cdot c\left(t\right) \, dt \tag{5.18}$$

$$J = \sum_{t=0}^{T} P\left(u\left(t\right)\right) \cdot c\left(t\right) \Delta t \tag{5.19}$$

where $P(u(t))$ represents the power consumption in terms of the control signal $u(t)$ and $c(t)$ represents the cost. Equation (5.18) reduces to Equation (5.19) when it is discretized. To incorporate the effects of time-of-use (TOU) electrical pricing, an extra term is added to the formulation as follows:

$$J = \sum_{t=0}^{t_f} P\left(u\left(t\right)\right) \cdot c(t)\Delta t + \begin{cases} c_{d,\text{on}} \cdot \max\left(P\left(u\left(t\right)\right)\right) \text{ if } t \in t_{\text{on}} \\ c_{d,\text{off}} \cdot \max\left(P\left(u\left(t\right)\right)\right) \text{ if } t \in t_{\text{off}} \end{cases} \tag{5.20}$$

where c_d refers to a constant for the demand charge during peak hours, t_{on} represents the peak hour, and t_{off} represents the off-peak hours. Finally, the optimization problem is described as Minimize J:

$$\overrightarrow{u(t)} \in \mathbb{R}^+ \tag{5.21}$$

$$\text{s.t.} \, 0 \le c\left(t\right) \le 1 \, \forall t \in t_f \tag{5.22}$$

$$u_c\left(t\right) + u_d\left(t\right) \le 1 \, \forall t \in t_f \tag{5.23}$$

Two baselines are defined for the testing of the MPC scheme — without TES and with TES operating with a fixed schedule. Additionally, two separate pricing schemes are also tested — TOU pricing and real-time pricing (RTP). In the former, fixed electrical tariffs are assumed for peak and off-peak hours. This allows for simpler cost computation as a whole. In the latter, continuously changing electrical tariffs are assumed. In this case, the pricing data are obtained from an industrial source [26] which features live energy pricing data, which is shown in Figure 5.26 for the period spanning 21st to 22nd

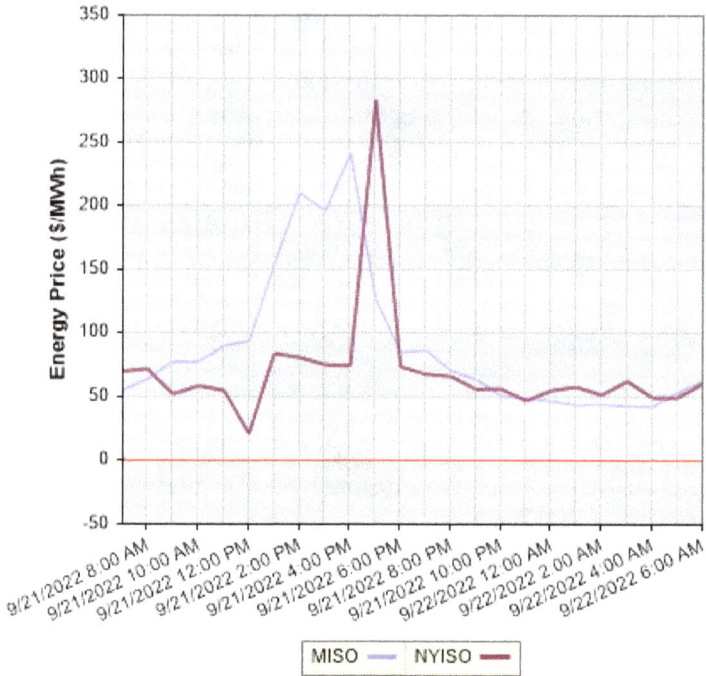

Figure 5.26.   Real-time energy pricing as reported by LCG consulting for the 21st–22nd September 2022 [26].

September 2022. Accordingly, the GA parameters are adjusted to enable a deeper search to be conducted into the space domain due to shorter time steps.

Under TOU pricing, the scheme was found to be able to generate significant cost savings of up to 15.9–16.4% for the without-TES scenario and 9–10% savings over the fixed-TES-operating-schedule scenario [14]. This was consistent for both the physical and NN modelling methods. Under RTP, the MPC using the physical modelling method was found to be able to similarly produce between 14% and 15% savings when compared to the one without TES and 8.2–10% savings when compared to one operating with a fixed TES schedule. Comparatively, the NN model recorded savings spanning 8–15%.

Figure 5.27 portrays the control signal fluctuations and the ice storage charging levels throughout a 24-h test period for both TOU and RTP.

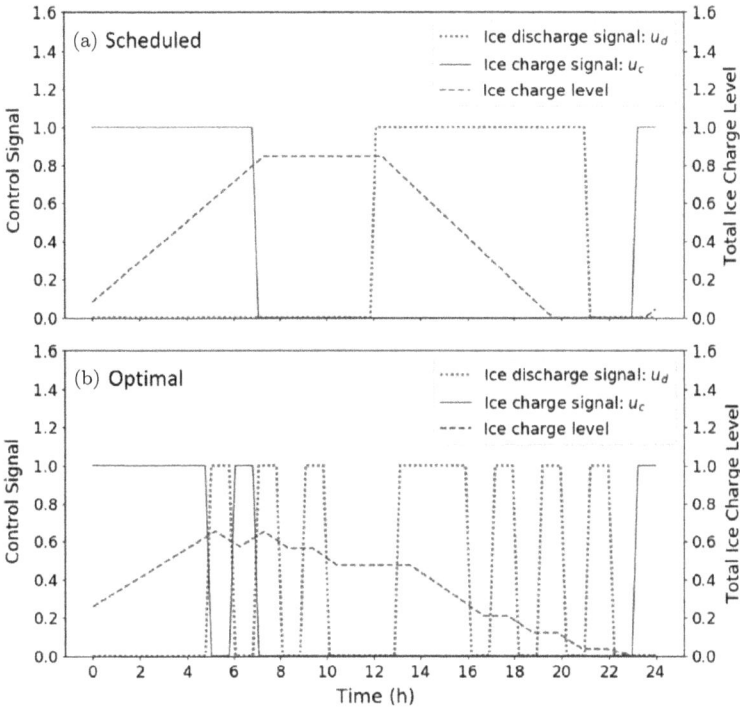

Figure 5.27. Ice storage control signals and total ice charge levels for TOU and RTP for 24-h discretization under (a) fixed TES operating schedule and (b) the proposed MPC scheme.

5.3. Conclusions

Understanding the process integration of TES is an important step in designing, sizing, and optimizing TES to achieve desired performance. As thermal systems around the world move progressively towards increased integration and decentralization, new knowledge and expertise are necessary to facilitate a more holistic and comprehensive understanding of the storage-integration framework.

In this chapter, the key fundamentals of process integration for TES and economic assessment methods are covered. Numerous recent examples involving TES integrated into district thermal systems are presented. Special attention has also been provided to recent research interest in cold recovery from LNG regasification. Finally, the subject of control systems and modern-day "smart" systems is also documented and illustrated with numerous case studies from the literature.

References

[1] D. Gibb, M. Johnson, J. Romani, J. Gasia, L. F. Cabeza and A. Seitz, Process integration of thermal energy storage systems — Evaluation methodology and case studies, *Applied Energy*, vol. 230, pp. 750–760, 2018.

[2] B. Habeebullah, Economic feasibility of thermal energy storage systems, *Energy and Buildings*, vol. 39, pp. 355–363, 2007.

[3] S. Sanaye and A. Shirazi, Thermo-economic optimization of an ice thermal energy storage system for air-conditioning applications, *Energy and Buildings*, vol. 60, pp. 100–109, 2013.

[4] M. Sebzali, B. Ameer and H. Hussain, Economic assessment of chilled water thermal storage and conventional air-conditioning systems, *Energy Procedia*, vol. 18, pp. 1485–1495, 2012.

[5] Baltimore Aircoil Company, TSU-M ICE CHILLER® thermal storage unit. Available: https://www.baltimoreaircoil.com/products/ice-thermal-storage/tsu-m-ice-chiller-thermal-storage-unit.

[6] G. Towler and R. Sinnott, *Chemical Engineering Design*, 2nd Edition, Oxford, UK: Elsevier, 2012.

[7] Y. Shao, K. Soh, Y. Wan, Z. Huang, M. Islam and K. Chua, Multi-objective optimization of a cryogenic cold energy recovery system for LNG regasification, *Energy Conversion and Management*, vol. 244, p. 114524, 2021.

[8] C. Rathgeber, S. Hiebler, E. Lävemann, P. Dolado, A. Lazaro, J. Gasia, A. de Gracia, L. Miro, L. F. Cabeza, A. König-Haagen, D. Brüggemann, A. Campos-Celador, E. Franquetf and B. Fumey, IEA SHC Task 42/ECES Annex 29 — A simple tool for the economic evaluation of thermal energy storages, in *International Conference on Solar Heating and Cooling for Buildings and Industry*, Istanbul, 2015.

[9] C. Rathgeber, E. Lävemann and A. Hauer, Economic top–down evaluation of the costs of energy storages — A simple economic truth in two equations, *Journal of Energy Storage*, vol. 2, pp. 43–46, 2015.

[10] K. Chua, S. Chou, W. Yang and J. Yan, Achieving better energy-efficient air conditioning — A review of technologies and strategies, *Applied Energy*, vol. 104, pp. 87–104, 2013.

[11] E. Guelpa and V. Verda, Thermal energy storage in district heating and cooling systems: A review, *Applied Energy*, vol. 252, p. 113474, 2019.

[12] W. Zhang, W. Hong and X. Jin, Research on performance and control strategy of multi-cold source district cooling system, *Energy*, vol. 239, p. 122057, 2022.

[13] A. Anderson, B. Rezaie and M. A. Rosen, An innovative approach to enhance sustainability of a district cooling system by adjusting cold thermal storage and chiller operation, *Energy*, vol. 214, p. 118949, 2021.

[14] S. J. Cox, D. Kim, H. Cho and P. Mago, Real time optimal control of district cooling system with thermal energy storage using neural networks, *Applied Energy*, vol. 238, pp. 466–480, 2019.

[15] D. Patel, J. Mak, D. Rivera and J. Angtuaco, LNG vaporizer selection based on site ambient conditions, GTI Energy, 2018. Available: https://www.gti.energy/wp-content/uploads/2018/12/Materials-4-Dhirav_Patel-LNG17-Poster.pdf.

[16] T. Yamamoto, Y. Fujiwara and S. Kitagaki, Challenges of advanced utilization of LNG cold in Osaka Gas Senboku LNG terminals, in *Design for Innovative Value towards a Sustainable Society*, Dordrecht, 2012.

[17] J. Khor, F. Dal Magro, T. Gundersen, J. Sze and A. Romagnoli, Recovery of cold energy from liquefied natural gas regasification: Applications beyond power cycles, *Energy Conversion and Management*, vol. 174, pp. 336–355, 2018.

[18] X. Peng, X. She, C. Li, Y. Luo, T. Zhang, Y. Li and Y. Ding, Liquid air energy storage flexibly coupled with LNG regasification for improving air liquefaction, *Applied Energy*, vol. 250, pp. 1190–1201, 2019.

[19] J.-Y. Ryu, A. Alford, G. Lewis, Y. Ding, Y. Li, A. Ahmad, H. Kim, S.-H. Park, J.-P. Park, S. Branch, S. Yu and C. Ryu, A novel liquid air energy storage system using a combination of sensible and latent heat storage, *Applied Thermal Engineering*, vol. 203, p. 117890, 2022.

[20] A. Tafone, E. Borri, L. F. Cabeza and A. Romagnoli, Innovative cryogenic Phase Change Material (PCM) based cold thermal energy storage for Liquid Air Energy Storage (LAES) — Numerical dynamic modelling and experimental study of a packed bed unit, *Applied Energy*, vol. 301, p. 117417, 2021.

[21] PCM Products Ltd., Sub-zero eutectic PCM solutions. Available: https://www.pcmproducts.net/Eutectic_Refrigeration_PCMs.htm.

[22] Y. Yau and B. Rismanchi, A review on cool thermal storage technologies and operating strategies, *Renewable and Sustainable Energy Reviews*, vol. 16, pp. 787–797, 2012.

[23] Hochschule Furtwangen University, Smart systems. Available: https://www.hs-furtwangen.de/en/research/research-themes//smart-systems/#:~:text=Smart%20Systems%20are%20%22intelligent%22%20devices,by%20means%20of%20an%20actuator.

[24] P. A. Østergaard, S. Werner, A. Dyrelund, H. Lund, A. Arabkoohsar, P. Sorknæs, O. Gudmundsson, J. E. Thorsen and B. V. Mathiesen, The four generations of district cooling — A categorization of the development in district cooling from origin to future prospect, *Energy*, vol. 235, p. 124098, 2022.

[25] Y. Xu, C. Yan, H. Liu, J. Wang, Z. Yang and Y. Jiang, Smart energy systems: A critical review on design and operation, *Sustainable Cities and Society*, vol. 62, p. 102369, 2020.

[26] LCG Consulting, Industry data. Available: http://www.energyonline.com/Data/Default.aspx.

Chapter 6

Thermal Energy Storage: Future Development and Roadmap

Abstract

As large-scale district cooling systems become more complex, the level of tediousness behind designing and applying TES systems increases. This is expected as the TES concept is largely centred on a static thermal capacitance used to balance mismatching supply and demand in thermal processes. Several key questions remain. First, how should the field continue to evolve to fit itself into the next-generation of thermal distribution landscapes? What directions would be necessary for TES research to stay relevant and applicable? This last chapter aims to provide some perspectives on these questions.

Keywords: Technology outlook; sustainability; Paris agreement; decarbonisation.

6.1. Thermal Energy Storage — The Present and Outlook

The International Renewable Energy Agency (IRENA) presents four different scenarios in its global renewable energy outlook for 2020 [1]:

(1) "Planned energy scenario", which outlines the situation based on current energy plans and currently established targets and policies.

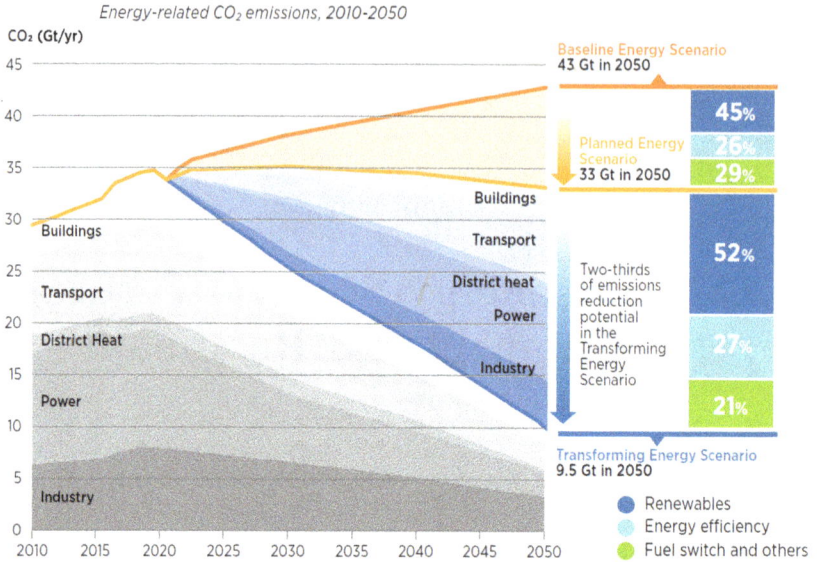

Figure 6.1. Annual energy-related CO_2 emissions by sector from 2010 to 2050 as projected by IRENA [1].

(2) "Transforming energy scenario", which describes a realizable stretch target scenario based on projected improved renewable energy sources and energy efficiencies, which would be minimally required for keeping the global temperature rise below 2°C as per the Paris Agreement [2].

(3) "Deeper decarbonization perspective", which expands the framework beyond the Transforming Energy Scenario, providing more novel views and choices for more rapid decarbonization, and suggests possibilities for accelerated action in specific fields.

(4) "Baseline energy scenario", which reflects the policies in place during 2015 — the time of the Paris Agreement.

The Transforming Energy Scenario outlines a reduction in the global economy energy intensity by about two-thirds, and in the trim energy-related CO_2 emissions by about 70%, approaching 2050, as illustrated in Figure 6.1. Further, the share of power generated via renewable sources needs to increase to around 86% from today's 26% [3].

Figure 6.2. The Vattenfall hot water TES currently under construction in Berlin, which is expected to be commissioned in early 2023 [4].

According to IRENA, a key aspect of this envisioned scenario is the significant reliance on variable renewable energy (VRE) sources such as solar photovoltaics and wind. In tandem, TES systems have bigger roles to play as they become critical pieces of the energy landscape, alongside electrical storage and demand-side management in providing flexibility for VRE sources [3]. From a transmission and distribution perspective, TES can also aid in mitigating the need for cost-intensive electricity network reinforcement. Enabling load-shifting in systems then leads to improved VRE utilization, grid congestion, and infrastructure investment. It is expected that by 2030, global TES capacity would grow threefold to almost 800 GWh, with investments of up to US$27.2 billion allocated to TES for power and cooling applications.

In urban contexts, STES systems such as chilled and hot water tanks are already highly established technologies in district thermal systems. One such example today is the colossal hot water tank currently under construction by *Vattenfall AB* in Berlin [4], shown in Figure 6.2. At 45 m tall and with a diameter of 43 m, the TES is capable of storing up to 56 million litres of hot water at 98°C and delivering as much as 200 MW of heat for 13 h. The location of the TES is conveniently situated next to a power-to-heat plant, which allows

Table 6.1.   Cost overview (in £) for TES technologies as reviewed by the British Department for Business, Energy, and Industrial Strategy (BEIS) in 2016 [5].

Type of TES (heat storage)	District heating	Non-domestic	Domestic
Tank TES	<£1–150/kWh (highly dependent on size, for example, some commercial systems may be similar in size to domestic)		£ 25–180/kWh
Pit TES	£0.30–0.80/kWh	N/A	N/A
Borehole TES	Potentially as low as £0.30/kWh (highly dependent on size and method used for measuring heat retained in ground)		
Aquifer TES	£600–1,000/kW (note ATES size commonly expressed as maximum heating rate for heat being extracted from the well, not the energy stored in the aquifer)		N/A
Latent TES	Unlikely to be used	£250–400/kWh (potentially as low as £50/kWh for large applications)	
Thermochemical TES	Potentially very cost-effective, but at current state of research, very cost-intensive and not ready or economical for commercialization		

conversion of VRE sources such as surplus wind or solar energy into heat stored in the TES.

Tank-based STES systems are easily deployed to meet short- to mid-term thermal needs. They perform well due to economies of scale that come with up-sizing thermal storage tanks, which see significant cost efficiencies at large industrial sizes in contrast to domestic or building-scale sizes. A review by the British Department for Business, Energy, and Industrial Strategy (BEIS) in 2016 [5] found capital costs for district-level tank-based STES installations to be between £1 and £150/kWh and as low as £0.3–0.8/kWh for long-term borehole-type and pit-type TES systems. Further, the cost of domestic-scale STES systems can be as high as £180/kWh, while the cost of LTES systems approaches £400/kWh as depicted in Table 6.1.

High-temperature TES for district heating constitutes the majority of the installed TES capacity worldwide. A significant portion of

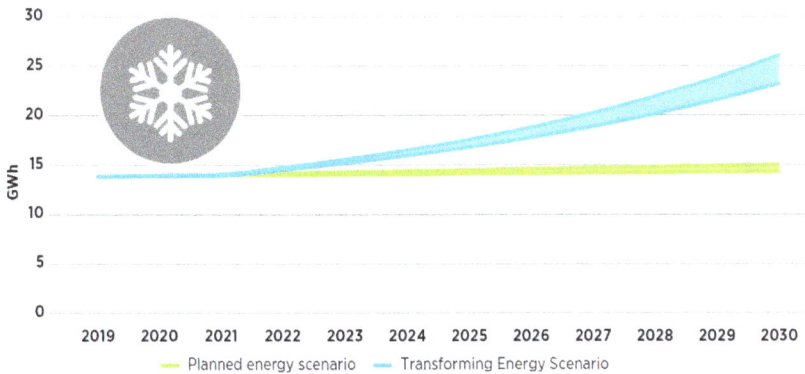

Figure 6.3. Installed and projected TES capacity for space cooling applications for both the planned and transforming energy scenario [3].

these systems includes long-term seasonal storages such as Aquifer and Borehole TES systems, which are crucial for temperate regions experiencing seasonal temperature changes. Contrastingly, tropical regions, experiencing high temperatures year-round, focus on cooling needs. This is evident from the 400 TES projects studied by IRENA in 2019, of which 160 are related to cooling in buildings and district cooling systems (DCS). The total figure is 13.9 GWh of installed TES capacity. As global temperatures continue to warm in the coming years due to climate change, this figure is expected to increase exponentially as shown in Figure 6.3 [3]. The planned energy scenario also predicts an increase in TES uptake, albeit a milder one. In sum, these data trends underline the importance of TES involvement to achieve desired climatic goals.

STES, such as chilled water storage tanks, constitute a significant fraction of the installed TES capacity in district cooling applications [5–7]. Ice-based [8,9] and other PCM-based LTES systems [10] are also seeing considerable uptake. DCS is capable of decoupling thermal demand and supply needs as well as providing high flexibility in storage over varying time scales and consumption patterns.

Figure 6.4 illustrates the deployment of an ice-storage system at Marina Bay Sands, Singapore, where the ice-based LTES is charged during periods of low electrical demand and discharged during high-demand periods. Currently, a pilot ice-based TES facility is being designed and engineered to increase this existing capacity by up to 3,000 refrigeration tonne-hours [9]. The total cooling capacity of the

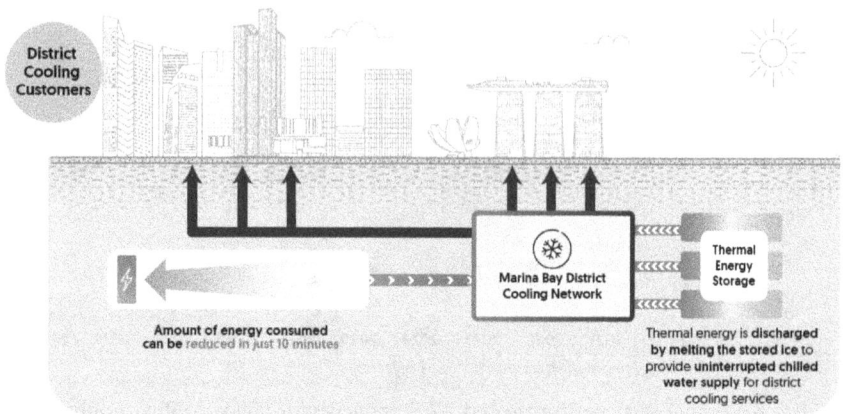

Figure 6.4.   Discharging the ice-based TES at the Marina Bay District Cooling Network in Singapore, which comprises up to 28 buildings [12].

district is anticipated to approach 73,000 refrigeration tonne-hours. The facility is expected to be completed by 2026 and trim annual CO_2 emissions by almost 20,000 tonnes. Based on present successful trials in Canada, China, and Europe, IRENA expects that future TES systems in DCS would involve more new developments in PCM and solid-state solutions. The following decade (5–10 years) of research into PCMs is, therefore, expected to be the primary driver for TES deployment in the DCS field. The projected storage efficiencies are expected to be 92% and above by 2030 [3].

At a smaller scale, PCM-based "thermal battery" TES units for domestic purposes [13,14] have been developed as shown in Figure 6.5. Live trials demonstrated cost savings of between 47% and 57% with reduced carbon emissions between 17% and 36%. Other examples of smaller-scale LTES units for transportation across long distances have also been trialled in the cold-chain sector. As elaborated in the work by Khor *et al.* [15], the use of low-temperature PCM LTES units for cold-storage in transport is a growing field of interest in the academic and industrial sectors. A real-life case study involved a collaboration between the University of Birmingham and a Chinese railway rolling stock company to develop a first-of-its-kind PCM-incorporated shipping container [16], as portrayed in Figure 6.6. Test results revealed that the containers were able to maintain at an interior condition of 5–12°C for up to 140 h, covering a total distance of 35,000 km by road and 1,000 km by rail.

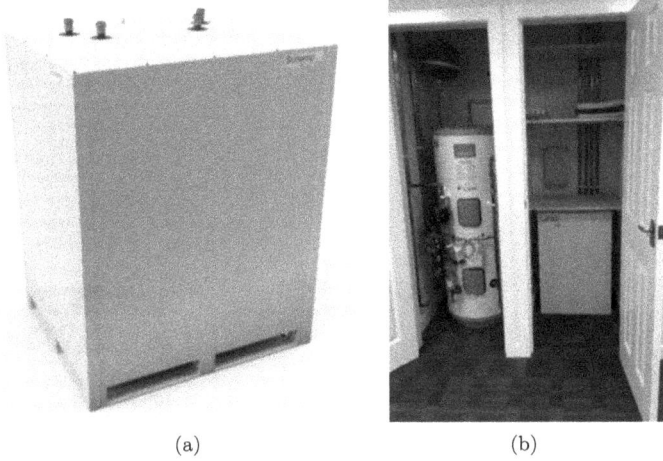

(a)   (b)

Figure 6.5.   PCM-based thermal battery (a) and its deployment in a domestic setting in comparison with a larger domestic storage heater (b) [13].

Figure 6.6.   The PCM-based refrigerated shipping containers developed by the collaboration between the University of Birmingham and CRRC Shijiazhuang [17].

One other key aspect of TES development is the incorporation of Smart control systems that facilitates demand-side management of the TES. One such example is Axiom Cloud [18], which developed a cloud-based "virtual battery" application to optimize cost

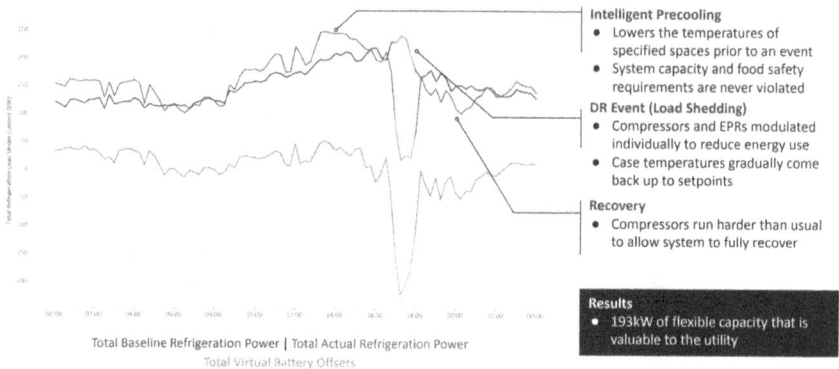

Figure 6.7.   The outcome of the North Californian trial by Axiom Cloud with Hello Fresh, a fresh food distribution company based in Berlin. The nature of this end user is such that refrigeration uptime and energy costs are of high priority [19].

savings and energy consumption based on time-changing electrical tariffs and user demand. The system has been trailed for supermarket and distribution centres equipped with PCM-based LTES, showing promising reductions in both peak demand and cost, as illustrated in Figure 6.7 [19].

It is largely expected that, in the short term, a large portion of development in the LTES sector for thermal distribution at building-level or district levels would lie in the development and availability of PCMs. For instance, lithium chloride for high-temperature storage can be substituted by sodium chloride for a fraction of the price ($250/kg vs. $20/kg). Low-temperature storage systems can branch out to utilizing PCMs beyond conventional ice-based LTES; leading to significant cost advantages [3]. However, corrosion and volume expansion issues of PCMs have to be judiciously considered during material selection in such installations. Research into mitigating solutions of such effects in operation becomes important in promoting their commercial viability in TES installations. One recent solution that has been developed to facilitate PCM selection for cold-chain food storage is the implementation of a FRISBEE[1] software project [20]. As illustrated in Figure 6.8, the software takes into

---

[1]Food Refrigeration Innovations for Safety, consumers' Benefit, Environmental impact, and Energy optimization along the cold chain in Europe.

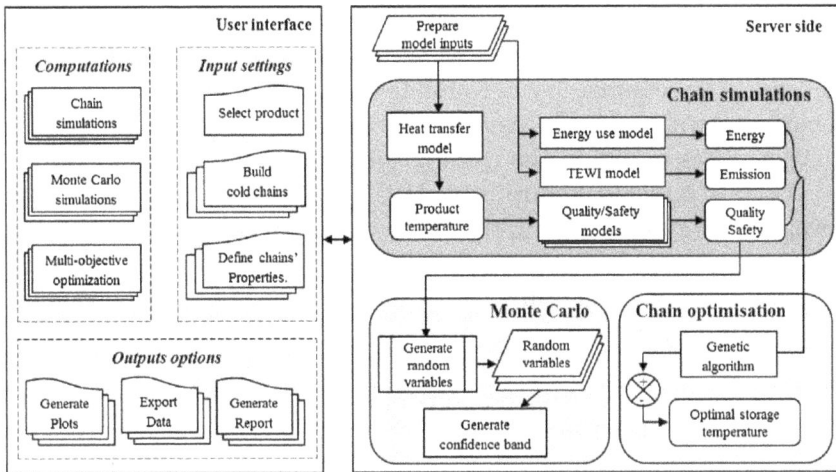

Figure 6.8. The software design for the FRISBEE tool.

account the specific application and assesses the most suitable PCMs based on the resulting product (food) quality, energy use, and CO_2 emissions. It employs validated kinetic storage models as well as a Monte Carlo simulation for optimization.

Thermochemical TES systems are also expected to command greater interest in the long term (10+ years), with pilot-scale demonstrations entering the fray around 2050. At present, the limited implementation of industrial-level thermochemical TES systems is due to high investment costs as well as concerns over material incompatibilities. The latter factor is attributed to the corrosiveness of certain sorption materials [5]. Thermochemical TES systems are likely to see greater adoption in applications where high thermal density and storage stability are key concerns. These applications may also include the integration of VRE sources, especially with improving smart technologies and integration methods [3].

One pilot-scale innovative integration of thermochemical TES and district thermal systems has been undertaken by the *H-DisNet* (Intelligent Hybrid Thermo-Chemical District Networks) [21]. As portrayed in Figure 6.9, the concept utilizes a hybrid thermal transport system that pumps partially hydrated or dehydrated salts around a district. The mobile thermochemical carrier facilitates remote utilization in different parts of the district. The key to this innovation lies in the core principles of sorption processes: absorption and desorption do not occur in the same place as both require different temperatures and partial pressures. By decoupling the absorption and desorption locations, as shown in Figure 6.9(b), a districtwide heat pump system that operates based on the moving sorption fluid can be realized.

As illustrated in Figure 6.10, IRENA highlights the application level of each type of TES in the district heating and cooling domains. Tank-based and underground STES systems are already well-established technologies in TES applications. For LTES deployments, ice-based systems remain the most commonly employed. Comparatively, high-temperature PCMs for heating are still in their demonstration phase, while sub-zero low-temperature/cryogenic PCMs are at prototyping level. Key barriers to a greater prevalence of these types of LTES lie in the inherently undesirable material properties such as corrosion and hysteresis. Thermochemical TES systems remain costly despite the high application expectations. While most projects to date are at pilot-scale, higher adoption is expected in the long term when PCM technology advances and greater system-level integration may be achieved with higher system intelligence.

6.2. Thermal Energy Storage — Future Research and Trends

Key TES research involves either material-level research or system-level optimization and process integration, the latter of which has been covered in detail in Chapter 5. In material development, one interesting field that has seen increased attention in recent years is the use of concrete as a thermal storage material. The use of concrete in building materials enables structural reinforcement as well as indoor comfort arising from its high thermal capacity [22]. Boquera *et al.* [22] observed that even though high-temperature storage systems employ molten salts, ceramic materials such as alumina-silicate geopolymers and refractory bricks can also be used. Liu and Yang [23], for instance, conducted numerical simulations as well as an optimization on a shell-and-tube type concrete-based TES unit as portrayed in Figure 6.11(b). Martelleto *et al.* [24] incorporated PCM to concrete in a similar configuration as shown in Figure'6.11(a) and achieved almost double the amount of thermal storage.

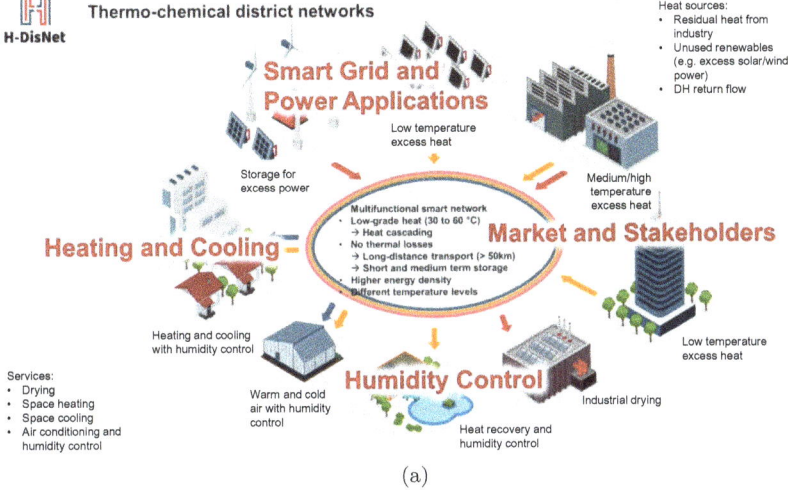

(a)

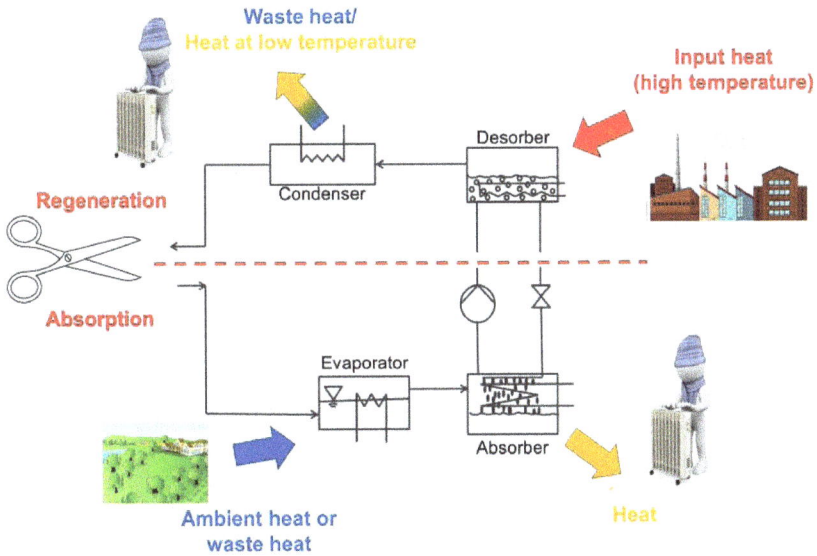

(b)

Figure 6.9. (a) Key features and (b) schematic representation of the operating flow of the H-DisNet project [21].

District heating and cooling

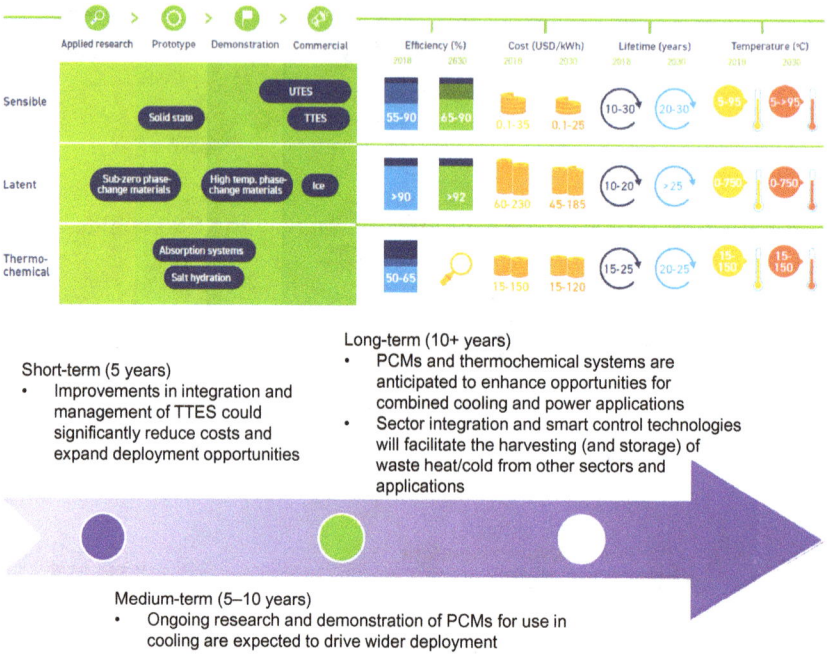

Figure 6.10.　Projected timeline for TES deployment for district heating and cooling using sensible, latent, and thermochemical storage systems. TTES refers to tank thermal energy storage and UTES refers to underground thermal energy storage [3].

Mikkelsson *et al.* [25,26] extended the study of such hybrid TES to system-level analysis, assessing the impact of varying piping configurations as well as control methods on system-level performance.

At pilot-scale testing, a company from Colorado [27] has developed a concrete-based TES integrated with steam-based power plants to decouple fuel consumption from electrical generation. Three separate concrete mixtures with embedded tubes were cycled more than 1,500 times with steam between 400°C and 600°C over 5,000 h to evaluate any changes in their material properties. The selected mixtures were then deployed in a power plant in Alabama in collaboration with Storworks Power [27,28]. Testing is currently still underway, and the operating flow of the proposed TES-integrated system is illustrated in Figure 6.12. During off-peak hours, steam

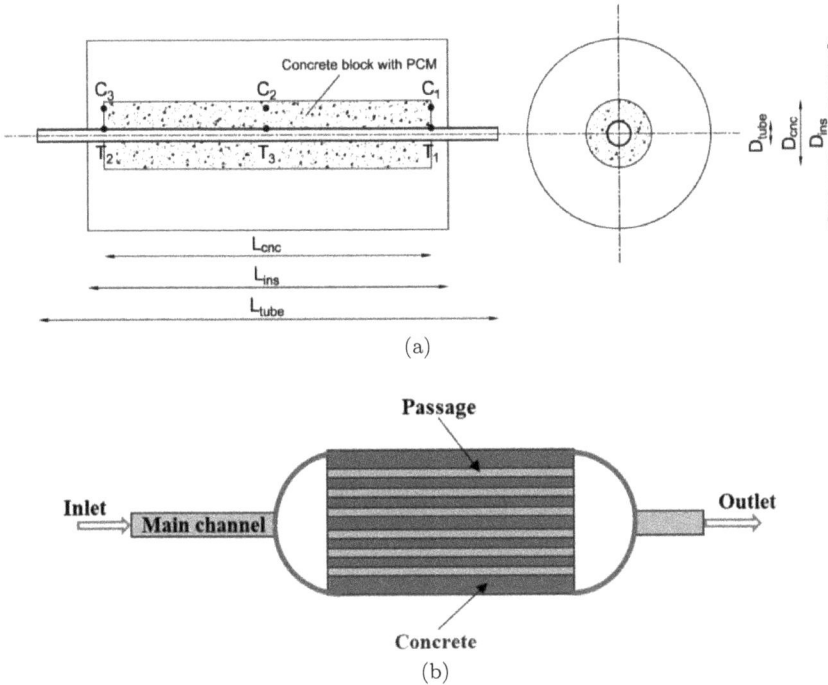

(a)

(b)

Figure 6.11.   Shell-and-tube type concrete TES units as studied by (a) Martelleto *et al.* [24] and (b) Liu and Yang [23].

from the unit is directed to the concrete TES unit to charge it, condensing the steam that passes through to produce water which is stored in a 12,000-gallon tank. During periods of higher demand, the water is fed back through the concrete TES to supply superheated steam to the turbine for additional power generation.

Boquera *et al.* [22] have identified some key research gaps in the field as follows:

(1) The need to consider carbon emissions footprint for the manufacture of cement during its life-cycle for system-level optimization.
(2) The acoustical aspect of the study should be incorporated as a complement to the central thermal of comfort that is under consideration.
(3) Greater attention should be paid to long-term deployment issues such as corrosion, thermal stability, and durability of the material.

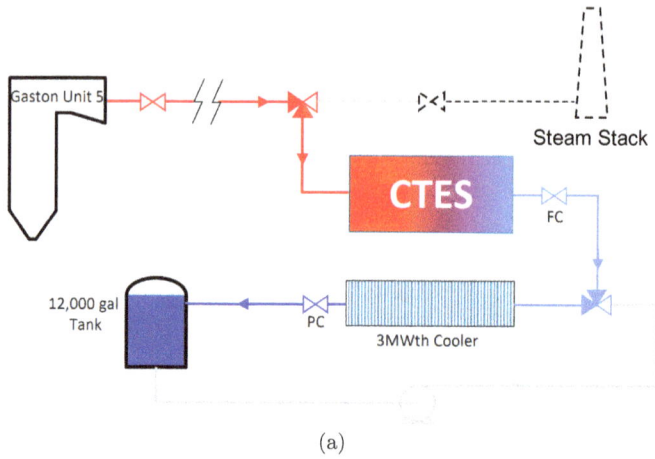

(a)

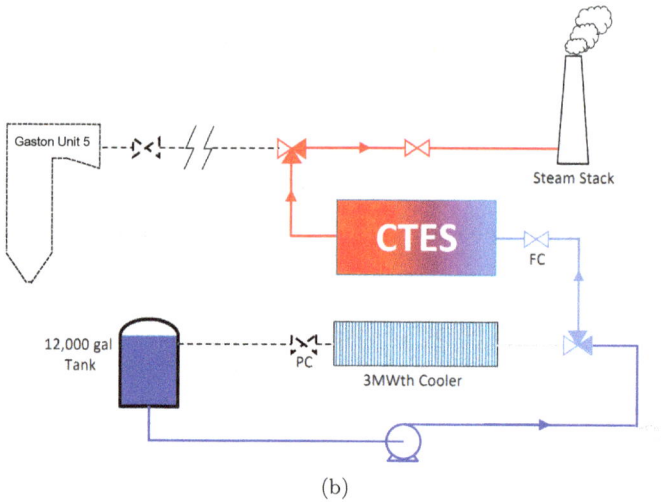

(b)

Figure 6.12. (a) Charging and (b) discharging modes of the concrete TES (CTES) unit as integrated with the Alabama Power Plant Gaston unit [28].

(4) More research activities should be devoted to the study of pore pressure control, which triggers pore fracture. Studies should also include investigating the effect of additional thermal resistance in heat-exchanger-type configurations on the layers of concrete.

(5) Incorporate other forms of thermal storage such as thermochemical materials, or the use of recycled waste materials for sustainability.

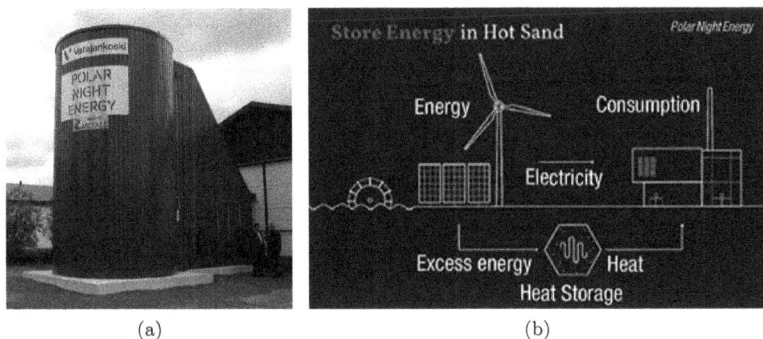

(a)   (b)

Figure 6.13.   (a) The sand-based TES [29,30] and (b) the schematic representation of its operation [31].

Interest in other less common thermal storage materials at pilot scales is also gaining traction. One recent example is the sand-based TES unit set up by a Finnish thermal storage company [29] and situated within a power plant operated by a local utility company.

The system utilizes heated sand at 500–600°C in a 4 m (D) × 7 m (H) cylindrical storage tank as shown in Figure 6.13. With a total storage capacity of 8 MWh, the storage is heated via VRE sources and can discharge up to 100 kW of heat over 80 h. The operations of these TES units are crucial during times of power shortage, such as the present scenario where the supply of natural gas from Russia to Finland has been discontinued due to the ongoing military conflict.

Recently, a similar commercial STES trial utilizing graphite was also set up in *Wodonga*, Australia [32]. Similar to the Finnish sand-based TES, the graphite-TES utilizes VRE from the grid to heat graphite blocks to up to 900°C as depicted in Figure 6.14(a). During the discharging, water is pumped through the storage to be superheated to 150–200°C steam, which is then supplied to a pet-food factory for cooking and supporting its canning process, as portrayed in Figure 6.14(b). The system has reduced gas consumption by up to 20%, with a 100% VRE reliance by 2040 [32]. The system was installed in Lake Cargelligo in Central NSW, Australia [32].

Many commercial pilot installations have primarily utilized high temperatures to overcome the low sensible heat capacity of their corresponding storage materials. Other endeavours have utilized latent heat instead. For example, one company has taken the sand-based

(a)　　　　　　　　　　　　　　　　(b)

Figure 6.14.　(a) The graphite-based TES by Graphite Energy [33] and (b) the pet-food factory where steam is supplied [32].

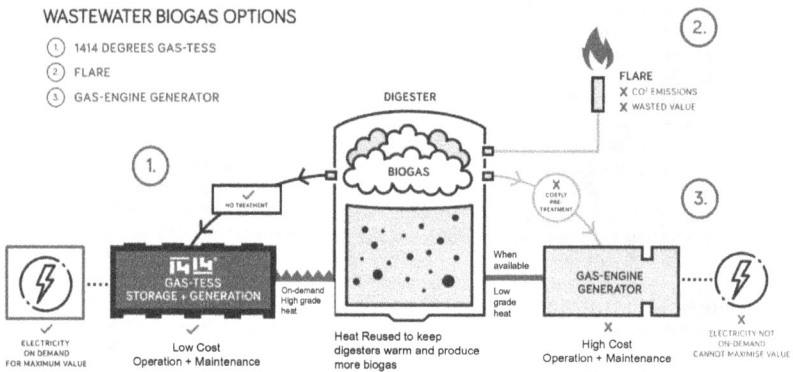

Figure 6.15.　The biogas-integrated TES system at Glenelg, South Australia [35,36].

TES concept further by introducing a molten silicon-based LTES that melts at the ultra-high temperature of 1,414°C. It also employed VRE and waste-heat sources for charging and supplying heat and power to end users [34]. The company has undertaken several key projects in South Australia, and notably set up a biogas-integrated TES system [35] as shown in Figure 6.15.

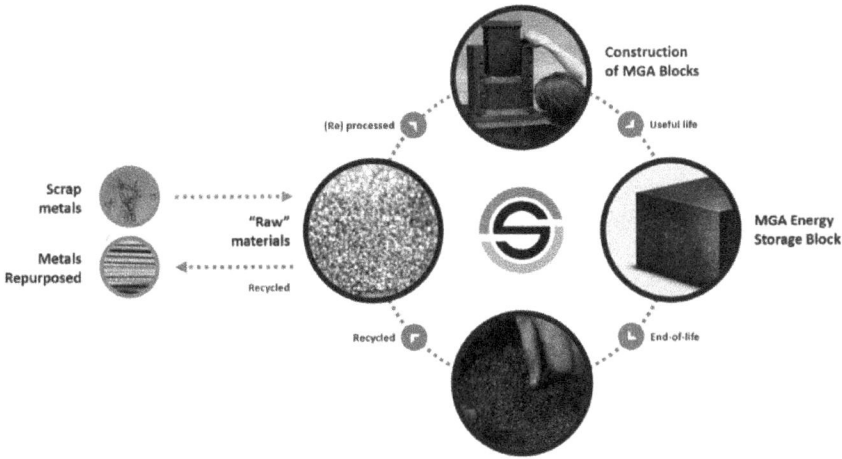

Figure 6.16. Production of the form-stabilized composite energy storage block [37].

In wastewater treatment plants, biogas that is produced can be combusted to support CHP processes. However, the gas requires pre-treatment prior to combustion and cannot be stored for later use. This leads to excess availabilities at times when demand is low. Integrating the biogas system with a TES unit allows the biogas to be instantly combusted to produce heat or electrical energy for heat, which is then stored in the TES. Accordingly, a constant supply of high-grade heat to the wastewater treatment plant is provided, while also ensuring that the biogas is not excessively combusted or wastefully exhausted to the atmosphere. The facility at *Glenelg*, Australia, as schematically illustrated in Figure 6.15, allows the TES to supply low-grade heat in the form of hot water around 60°C for the plant's digestion process. Additionally, the heat stored is used to produce electricity via steam at around 200°C [36].

Another Australian research group, MGA Thermal [37], has taken a novel approach to the high-temperature LTES concept by developing a form-stable composite PCM. The LTES unit comprises scrap metal PCM particles dispersed within a proprietary porous matrix, as shown in Figure 6.16. Present prototypes have shown high-temperature thermal storage potentials ranging between 600°C and 800°C.

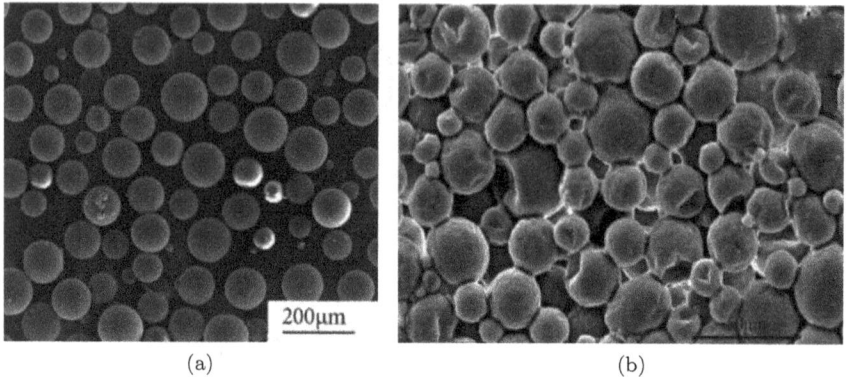

(a)   (b)

Figure 6.17.   SEM micrographs of a PMMA shell encapsulating n-hexadecane (a) [39] and PU shell with butyl stearate core (b) [40].

One niche area of PCM form stabilization that is receiving wider attention is microencapsulation. Macroencapsulation involves the employment of larger-scale "shells" such as spheres, tubes, and slabs. A review by Cardenas-Ramirez *et al.* [38] highlights that PCM microencapsulation research to date has been focused on the structural properties of the shell material as well as the encapsulation methods. Microencapsulation shell materials are broadly classified into organic and inorganic. Organic shell materials such as urea-formaldehyde (UF) and melamine-formaldehyde (MF) are known to produce toxic compounds while possessing poor thermal properties. Thus, research has shifted to less-harmful polymers such as polystyrene (PS) and polymethyl methacrylate (PMMA), as shown in Figure 6.17.

Inorganic shell materials such as silica, calcium carbonate, and titania have been tested for applicability in PCM microencapsulation. For instance, studies conducted by Yu *et al.* [41] and He *et al.* [42] reported that the calcium carbonate shell precipitated using the self-assembly method, as depicted in Figure 6.18(b), lost up to 38% of the core material over 30 days. The precursors used were calcium chloride and sodium carbonate, and the shell was precipitated onto the PCM, which was n-octadecane.

As the technology is largely unexplored beyond lab scales, few known players in the commercial space exist. A review of the market, however, did reveal at least one commercial attempt at utilizing microencapsulated PCMs in textiles, beddings, and medical device

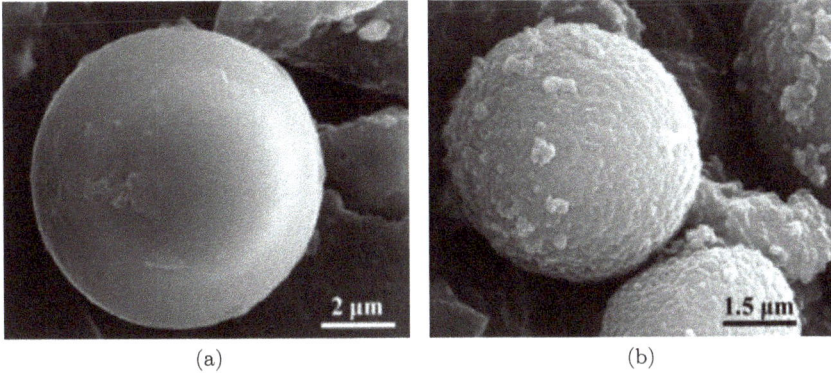

Figure 6.18. SEM micrographs of the *n*-octadecane PCM encapsulated by (a) sodium silicate [41] and (b) calcium chloride [42].

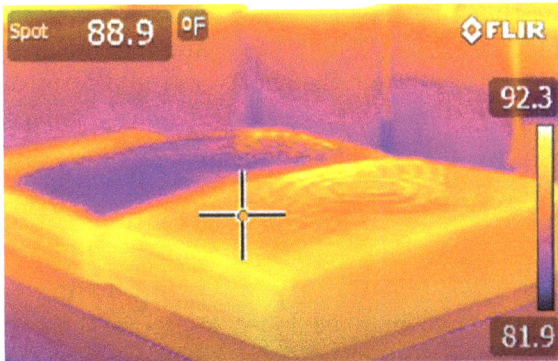

Figure 6.19. Test comparison between a standard mattress and a mattress incorporated with microencapsulated PCM [43].

applications. Figure 6.19 shows a side-by-side comparison between one of these products claiming to retain cooler temperatures for longer periods of time compared to a control mattress sample.

The limited uptake of this thrust of PCM technology at commercial scale is understandable — many studies have neither addressed the thermal transfer properties nor the overarching thermal performance of the LTES at macro-scale as noted by Cárdenas-Ramírez *et al.* [38]. A progressive shift in focus is hence imperative for this innovation to be eventually implemented in industry.

6.3.   Conclusion

As the development of climate change mitigating plans becomes critical, new R&D innovations on TES systems need to be proposed. TES systems play a central role in balancing out the instabilities and inconsistencies that arise from both society and the environment. Several key points that are worthy to note regarding the present and future progress of TES include the following:

(1) Cooling demand will only increase in years to come due to the increasing global temperatures. The focus on high-capacity TES for cold-chain and district-cooling applications will be key for both short- and long-term developments in global energy and carbon emissions management.
(2) It is clear that higher adoption rates of latent TES systems at industrial scale are important to enhance the downstream energy efficiency holistically for the short- and mid-term outlooks. For sensible TES systems, improved process-level management translates to measurable efficiency improvements.
(3) Thermochemical TES systems hold promising possibilities for the long term but still require significant research and test-bedding in live situations. Similarly, other innovations such as microencapsulated and composite/form-stabilized PCM will require further proof-of-concept testing at pilot scales before their true potential at the large-scale system level can be assessed.
(4) Advancement in TES research cannot be viewed as distinct and separate from thermal distribution systems. Technological advancement at the fundamental level only is the primary driver for TES implementation. Its potential cannot be fully realized without furthering our understanding of its role when it is integrated into a system.

References

[1] IRENA, Global renewables outlook: Energy transformation 2050, International Renewable Energy Agency, 2020.
[2] UNFCCC secretariat, The Paris Agreement. Available: https://unfccc. int/process-and-meetings/the-paris-agreement/the-paris-agreement.

[3] IRENA, Innovation outlook: Thermal energy storage, International Renewable Energy Agency, Abu Dhabi, 2020.

[4] Vattenfall AB, Germany's largest heat storage in the starting blocks, Vattenfall AB, 13 July 2022. Available: https://group.vattenfall.com/press-and-media/newsroom/2022/germanys-largest-heat-storage-in-the-starting-blocks.

[5] Department of Business, Energy and Industrial Strategy (BEIS), Evidence gathering: Thermal energy storage (TES) technologies, London, 2016.

[6] ENGIE, District cooling providing energy to communities through efficient and decarbonised district energy solutions, 2021. Available: https://www.engie-sea.com/district-cooling.

[7] SP Group; Temasek Holdings, *Taking the Heat off Cooling: A Greener Way to Cool*. SP Group; Temasek Holdings, Singapore, 2021.

[8] Keppel Corporation Limited, Singapore Plants, 2010. Available: https://www.keppeldhcs.com/singapore_plants.html.

[9] Pendinginan Megajana Sdn. Bhd., District cooling, 2021. Available: http://megajana.com.my/megajana/district-cooling/.

[10] C. Tan, New energy storage system at George Street Substation to help sustainably cool the Marina Bay area, *The Straits Times*, 29 August 2022. Available: https://www.straitstimes.com/singapore/new-energy-storage-system-at-george-street-substation-to-help-cool-buildings-in-marina-bay-area.

[11] Keppel Corp, New technology to boost energy efficiency of district cooling systems, 14 October 2021. Available: https://www.ema.gov.sg/cmsmedia/News/Media%20Release/2021/141021-Media-Release-New-Technology-to-Boost-Energy-Efficiency-of-District-Cooling-Systems.pdf.

[12] SP Group, How does thermal energy storage system support demand response?, SP Group. Available: spgroup.com.sg/wcm/connect/spgrp/313c77a8-d6cd-41d8-a8f9-15f5b3643a82/Infographic+-+Thermal+Energy+Storage+and+Demand+Response.pdf?MOD=AJPERES&CVID=.

[13] Sunamp, An introduction to thermal energy storage, 2019.

[14] Sunamp, Sunamp Global. Available: https://sunamp.com/#home/.

[15] J. Khor, F. Dal Magro, T. Gundersen, J. Sze and A. Romagnoli, Recovery of cold energy from liquefied natural gas regasification: Applications beyond power cycles, *Energy Conversion and Management*, vol. 174, pp. 336–355, 2018.

[16] University of Birmingham, Passively cooled containers being delivered for integrated rail and road cold chain transportation following world's first commercial demonstration, University of Birmingham,

7 August 2019. Available: https://www.birmingham.ac.uk/news/2019/passively-cooled-containers-being-delivered-for-integrated-rail-and-road-cold-chain-transportation-following-worlds-first-commercial-demonstration.

[17] University of Birmingham, UK and China scientists develop world-first cold storage road/rail container, University of Birmingham, 21 December 2018. Available: https://www.birmingham.ac.uk/news/2018/uk-and-china-scientists-develop-world-first-cold-storage-roadrail-container.

[18] Axiom Cloud, Monetize refrigeration management with Virtual Battery, Axiom Cloud,. Available: https://www.axiomcloud.ai/virtual-battery.

[19] Slipstream Inc, Case studies in energy storage systems for refrigeration, 18 July 2022. Available: https://slipstreaminc.org/sites/default/files/events/rtes-case-studies-webinar-slides.pdf.

[20] S. Gwanpua, P. Verboven, D. Leducq, T. Brown, B. Verlinden, E. Bekele, W. Aregawi, J. Evans, A. Foster, S. Duret, H. Hoang, S. van der Sluis, E. Wissink, L. Hendriksen, P. Taoukis, E. Gogou, V. Stahl, M. El Jabri and A. Geeraerd, The FRISBEE tool, a software for optimising the trade-off between food quality, energy use, and global warming impact of cold chains, *Journal of Food Engineering*, vol. 148, pp. 2–12, 2015.

[21] H-DisNet, PROJECT H-DisNet, 2016. Available: https://www.h-disnet.eu/.

[22] L. Boquera, J. R. Castro, A. L. Pisello and L. F. Cabeza, Research progress and trends on the use of concrete as thermal energy storage material through bibliometric analysis, *Journal of Energy Storage*, vol. 38, p. 102562, 2021.

[23] C. Liu and H. Yang, Multi-objective optimization of a concrete thermal energy storage system based on response surface methodology, *Applied Thermal Engineering*, vol. 202, p. 117847, 2022.

[24] F. Martelleto, L. Doretti and S. Mancin, Numerical simulation through experimental validation of latent and sensible concrete thermal energy storage system, *Journal of Energy Storage*, vol. 51, p. 104567, 2022.

[25] D. Mikkelson and K. Frick, Analysis of controls for integrated energy storage system in energy arbitrage configuration with concrete thermal energy storage, *Applied Energy*, vol. 313, p. 118800, 2022.

[26] D. Mikkelson and J. M. Doster, Investigation of two concrete thermal energy storage system configurations for continuous power production, *Journal of Energy Storage*, vol. 51, p. 104387, 2022.

[27] M. Matz, A new use for a 3,000-year-old technology: Concrete thermal energy storage, *EPRI Journal*, 19 May 2021. Available: https://

eprijournal.com/a-new-use-for-a-3000-year-old-technology-concrete-th ermal-energy-storage/.

[28] S. Hume, Concrete thermal energy storage enabling flexible operation without coal plant cycling, Electric Power Research Institute, 2021.

[29] Polar Night Energy, Sand battery. Available: https://polarnight energy.fi/sand-battery.

[30] C. Murray, World's first large-scale "sand battery" goes online in Finland, Solar Media Ltd, 6 July 2022. Available: https://www. energy-storage.news/worlds-first-large-scale-sand-battery-goes-online-in-finland/.

[31] G. Ondrey, Commercial debut for sand-based thermal-energy storage, Access Intelligence, LLC, 1 August 2022. Available: https://www. chemengonline.com/thermal-energy-storage/?printmode=1.

[32] J. Purtill, A "graphite battery" in Wodonga will be Australia's first commercial thermal energy storage, *ABC News*, 4 August 2022. Available: https://www.abc.net.au/news/2022-08-04/graphite-battery-will-be-first-commercial-thermal-energy-storage/101295350.

[33] Graphite Energy, Proven, reliable graphite thermal energy storage, Graphite Energy. Available: https://www.graphiteenergy.com/ technology.

[34] 1414 Degrees, What is 1414 degrees? Available: https://1414degrees. com.au/what/.

[35] 1414 Degrees, GAS-TESS, 1414 Degrees. Available: https:// 1414degrees.com.au/gas-tess/.

[36] J. Parham, P. Vrettos and N. Levinson, Commercialisation of ultra-high temperature energy storage applications: The 1414 degrees approach, in *Ultra-High Temperature Thermal Energy Storage, Transfer and Conversion*, Woodhead Publishing, 2021, pp. 331–346.

[37] MGA Thermal, Introducing the MGA block, MGA Thermal. Available: https://mgathermal.com/tech.

[38] C. Cárdenas-Ramírez, F. Jaramillo and M. Gomez, Systematic review of encapsulation and shape-stabilization of phase change materials, *Journal of Energy Storage*, vol. 30, p. 101495, 2020.

[39] S. Lashgari, H. Arabi, A. Madhavian and V. Ambrogi, Thermal and morphological studies on novel PCM microcapsules containing n-hexadecane as the core in a flexible shell, *Applied Energy*, vol. 190, pp. 612–622, 2017.

[40] S. Lu, T. Shen, J. Xing, Q. Song, J. Shao, J. Zhang and C. Xin, Preparation and characterization of cross-linked polyurethane shell microencapsulated phase change materials by interfacial polymerization, *Materials Letters*, vol. 211, pp. 36–39, 2018.

[41] S. Yu, X. Wang and D. Wu, Microencapsulation of n-octadecane phase change material with calcium carbonate shell for enhancement of thermal conductivity and serving durability: Synthesis, microstructure, and performance evaluation, *Applied Energy*, vol. 114, pp. 632–643, 2014.

[42] F. He, X. Wang and D. Wu, New approach for sol–gel synthesis of microencapsulated n-octadecane phase change material with silica wall using sodium silicate precursor, *Energy*, vol. 67, pp. 223–233, 2014.

[43] Encapsys LLC, We're innovators who push the limits of excellence through microencapsulation. Available: https://www.encapsys.com/about/.

[44] Encapsys, EnFinit PCM 28 — Superior thermal management, Encapsys, 28 March 2016. Available: https://www.youtube.com/watch?v=vqcxEWv4J4M&ab_channel=Encapsys.

Index

www.ingramcontent.com/pod-product-compliance
Lightning Source LLC
Chambersburg PA
CBHW050559190326
41458CB00007B/2105